41558

Submerged-arc

0415

D0299134

64337

RI

SUBMERGED-ARC
Welding

P T Houldcroft

ABINGTON PUBLISHING

Woodhead Publishing Ltd in association with The Welding Institute

Cambridge   England

Published by Abington Publishing,
Woodhead Publishing Ltd, Abington Hall, Abington,
Cambridge CB1 6AH, England

First published 1977
Second edition 1989

© Woodhead Publishing Ltd

British Library Cataloguing in Publication Data
A CIP catalogue record for this book is available from the British Library.

ISBN 1 85573 002 2

Designed, typeset and printed by Crampton & Sons Ltd, Sawston,
Cambridge CB2 4BQ, England

CONTENTS

Preface

The first edition of 'Submerged-arc welding' was based on a collection of papers presented at a state-of-the-art seminar held in 1977. In the foreword to this edition the Technical Director of the seminar, Dr S B Jones, commented that 'such a review was undoubtedly necessary since, in the forty years of its existence, the submerged-arc process had seen development and diversification unparalleled in any other arc welding technique.'

A second seminar was held in 1987 which revealed that some of the directions being pursued in 1977 were no longer regarded as being quite so promising. Although the potential advantages still exist the early interest in increasing deposition rate by hot and cold wire additions or by long stickout techniques and DC electrode negative (DCEN) had been tempered by practical difficulties on the shopfloor. The number of electrodes used is now usually one, often two, sometimes three, but except for specialised applications such as pipe welding rarely more than three. Current interest in increasing deposition rate centres on metal powder additions. Methods for increasing joint completion rates now include narrow gap techniques. Major improvements in mechanisation and process control have also become available and are being applied.

This edition of 'Submerged-arc welding' incorporates much of the material from the two seminars but it has been completely rewritten both to update and to eliminate repetition and also to incorporate other Welding Institute material. New sections have been added on powder additions and narrow gap welding.

The book is intended as an introduction to an important and highly productive welding process. It is also hoped that it may provide encouragement to those with some knowledge of the process to examine the less widely used variants, many of which offer advantages in particular applications.

The editor wishes to express his indebtedness to Dr R B G Yeo, N Bailey and O K Gorton who made helpful comments on the manuscript and to those who took part in the seminars and provided the information forming the basis of this book. They were: Dr G Almquist, Dr-Ing B K Bersh, H C Buckingham, D J Ellis, I Gronbeck, A M Horsfield, Dr R L Jones, Dr S B Jones, N A Kennedy, V G Marshal, K N Middleton, G K Raven, D E H Reynolds, Dr C E Thornton, S R Usher and R T Wolfenden.

Permission to use illustrations is gratefully acknowledged to ESAB, Lincoln Weldro Ltd, F Bode and Son, Fairfield Mabey, Harland and Wolff plc, *etc, etc.*

P T Houldcroft
1989

CHAPTER 1
Origins and development

The process

The submerged-arc process is a well established and extremely versatile method of welding. Basically a simple process, in which an arc is struck from a continuously fed electrode wire to the work under a blanket of powdered flux, it was the first successful method of automatic arc welding.

The flux covering totally conceals the arc, preventing flash and glare, while smoke and fumes are almost absent. As it is impossible to see the arc or the molten pool simple devices are required to assist positioning of the welding head. The flux protects the arc and weld metal from the atmosphere and may also supply, through reactions between it and the molten metal, any additional elements needed to control porosity and provide the weld with the required mechanical properties. Other characteristics of the flux, such as its state of division and method of manufacture, influence its properties and the conditions under which it can be used. The flux also affects the contour of the final weld bead as it forms molten slag which rises to the surface of the molten pool and shapes the weld bead as it solidifies. Well made submerged-arc welds are noted for their smooth contours.

An electrical contact near to the arcing point, through which the wire slides, allows introduction of the welding current. Electrode wire for welding ferrous materials normally contains alloy additions such as manganese and silicon which, together with elements provided by the flux, give control of properties and quality. Because flux and wire are added separately the user has, if necessary, an extra dimension of control which is not possible with manual metal arc (MMA) welding in which the wire and flux combination are decided by the manufacturer.

The closeness of the electrical contact and the wire feeding mechanism to the arc allow use of wire diameters as large as 6mm, although small diameters of less than 1mm, similar to those employed in metal inert gas

welding can also be used. A correspondingly wide range of welding
ints is possible allowing metal with thicknesses from as little as 3mm
ɔ 100mm and more to be welded. Submerged-arc welds can be deeply
netrating and the process gives high joint completion rates. Although it
is usually associated with welding thick steel for applications such as
pressure vessel manufacture, shipbuilding and structural engineering,
the process also has an important role in high speed automatic welding of
smaller, thinner components. The welding equipment can be stationary
over moving work or it may be mounted on a tractor or gantry to traverse
the workpiece. In Fig.1.1 a submerged-arc welding head on a column and
boom is shown making a circumferential weld in a heavy fabrication, but

1.1 *Submerged-arc welding a circumferential joint in a large
fabrication. The welding head is mounted on a column and boom
and is an example of what is regarded as typical of the process
which, however, is capable of a great variety of work.*

the process is capable of use over a wide range of applications, both thick and thin.

Although it was a process conceived for automatic welding of the heavier sections of steel plate a semi-automatic version has been in use for many decades and was probably the first semi-automatic arc welding system in wide use. Originally in the semi-automatic version flux was dispensed from a small hopper mounted on the hand gun through which the wire was fed but mechanised flux feeding from a remote supply has now made this unnecessary and has increased the mobility of the process.

The flexibility of submerged-arc welding led quickly to the development of variants in which more than one wire was used. Different electrical connections were found to give different types of penetration bead and changes in electrode extension were used to alter burn-off rate. The range of applications quickly became extensive and included not just welding but also surfacing. The process is widely used to weld ferrous and stainless steels and has been applied to other alloys of the copper based type and even titanium.

To summarise, submerged-arc welding is a productive process capable of producing quality welds in a wide range of thicknesses in ferrous and stainless steels and even some non-ferrous metals. As the arc and molten pool are totally concealed beneath the flux there is no radiation from the arc and smoke and fume are minimal. The process exists in many forms, semi-automatic and automatic, for welding and surfacing. One of its few limitations is that all welding must be carried out in horizontal/vertical, flat or near flat welding positions.

The first twenty years

In his 1959 Adams Lecture C E Jackson traced the origin of the submerged-arc process to 1935 when Kennedy, Jones and Rodermund applied for a US patent covering electric arc welding under a blanket of granulated welding flux which completely submerged the welding zone. In a short time the process was firmly established in the form shown in Fig.1.2 and was being used at welding currents of about 1000A. Two pass and one sided (using copper backing bar) welding procedures were used. The high deposition rates possible using a single wire were exploited to the extent that AC welding currents of 4000A were claimed when using 12mm diameter rods as filler wires. Typical welding procedures used in the period around 1940 may be found in the AWS 'Welding handbook' for 1942. The benefits of series arc, parallel welding and multipower

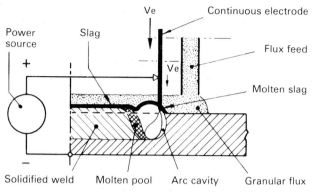

1.2 *Submerged-arc welding.*

welding* were established around 1950 and the techniques were used under production conditions, as was long stickout or I^2R heating for which deposition rates of 55 kg/hr were claimed.

The submerged-arc process was rapidly adopted for applications using carbon-manganese, low and high alloy steels. Comprehensive mechanical test data were available by the early 1940s giving weld metal tensile properties. Impact values (both Izod and Charpy V, mainly at room temperature) were quoted occasionally but there was little emphasis on heat affected zone (HAZ) properties. Non-destructive testing (NDT) was most often carried out by simple visual checks and X-ray examinations.

Electrode wires of the carbon-manganese and low alloy types were often similar to those in current use, and were therefore readily available. Where appropriate, wires of a matching composition to the plate material were also used. Submerged-arc welding was successfully used to weld ferritic and austenitic stainless steels as well as nickel, Monel, Everdur, and other non-ferrous materials.

The early fluxes were made by fusing the raw materials, usually in an electric furnace, and crushing and grading the product. Later, agglomerated fluxes were introduced which were made by mixing the constituents in the form of fine powders and then treating the mixture by pelletising and bonding or sintering the pellets in kilns. Sintering involves fritting but not fusing of the constituents.

---

*These variants of the submerged-arc process are discussed in Chapter 3. Series arc uses two wires with the two arcs in series from one power source; parallel welding uses two wires fed through the same contact tip giving two arcs in parallel from one power source; multipower welding uses several wires each with its own power supply.

The fluxes, both fused and agglomerated, which appeared to satisfy all the metallurgical and welding requirements at the time were of three main types:

1 High silica, with silica content around 50% and allowing the high welding currents which were then favoured. Some fluxes of this type are still in use.

2 Neutral, with silica content around 30%, these being generally preferred for multipass welding applications such as for pressure vessels. Fluxes of this type are also still in use.

3 Manganese-silicate, general purpose welding fluxes suitable for butt and fillet welding in the flat and horizontal/vertical positions.

These fluxes were used for high speed welding and were tolerant of poor plate conditions, *e.g.* rust. This type is still a significant proportion of the fluxes currently in use.

The process was used initially for applications which were relatively unimportant, *e.g.* for water pipe production. However, it soon gained acceptance for many of the applications which are now currently well established such as:

— pipe welding;
— shipbuilding;
— bridge building;
— pressure vessel construction;
— rebuilding and surfacing;
— welding in the motor industry;
— general fabrication.

At the end of the first twenty years both wires and fluxes were adequate to meet the requirements for the materials being welded at the time. The high deposition characteristics of the process were being exploited to the maximum within the known process knowledge. Although most applications employed single wires, multiwire processes were being used increasingly to extend the range of applications.

Development after 1955

Equipment and consumables

During the first 20 years the USA and Russia were responsible for many new developments but in the period after 1955 Japan and to a lesser extent Europe became process innovators. Multiwire welding was consolidated after 1955 and considerable refinements were made to the welding equipment. The special features of AC and DC (both DCEP and DCEN) were studied and understood and in the mid 1970s four wire systems were in use in the shipbuilding industry. At about the same time the six wire multiweave system was introduced for cladding applications such as lining nuclear pressure vessels.

Strip cladding using single and double strips instead of wire for the electrode became established. Additions of metal powder, cut wire, and continuous solid or cored wires were made to the molten pool to increase deposition rates and improve weld quality. Around 1970 the use of 'hot'* wires for butt welding and surfacing boosted deposition rates further while still maintaining high metallurgical quality in the deposited weld metal.

None of these innovations appeared to be limited by the equipment. Nevertheless deposition rates being obtained throughout industry were probably no higher than in the first twenty years. In the late 1970s and 1980s the submerged-arc process received much competition at the thin end of the market (up to 5mm thickness) from MIG and cored wire welding, however, the process remains widely used for light gauge applications such as domestic oil storage tanks and low pressure gas bottles. As a result of the development of metal powder addition techniques and narrow gap welding, use of the process for thicker material has been extended. For a while in the 1960s and early seventies electroslag welding was widely used in applications where submerged-arc welding might have been considered. Problems with HAZ properties of electroslag welds, which even now have not been resolved, have resulted in reduced use of electroslag for heavy welding and a resurgence of interest in submerged-arc welding.

Properties of joints

Weld metal property requirements in the first twenty years of the process were relatively easy to meet but limitations in the consumables being used subsequently became apparent. There began a trend for Charpy V impact testing, particularly below room temperature and at the same time grain-refined steels such as BS 968 and Lloyd's AH, DH and EH steels were introduced. Techniques using high welding currents and high dilution procedures which were previously successful were no longer satisfactory as the enhanced mechanical property requirements for the

---

*A 'hot' wire is a wire fed into the molten pool which carries enough current from a separate power supply to preheat the wire but not sufficient to fuse it and form an arc.

deposited weld metals could not be met. Flux compositions were examined in greater detail, particularly by the larger users. Whereas in 1942 it was sufficient to describe the submerged-arc welding flux as 'a specially granulated material of finely crushed mineral compositions usually gray or greenish gray in color', after 1960 this description was no longer acceptable. Much information became available on flux compositions and the effects of individual constituents on the running characteristics and mechanical properties. Studies of weld metal toughness properties using tests such as Charpy V, dropweight, bulge explosion, and CTOD, allowed wire and flux combinations to be classified more precisely on the basis of weld metal mechanical properties.

The various national specifications for submerged-arc fluxes now give more information but vary greatly in their approach. In the USA the wire is defined by the American Welding Society (AWS) specification, A5.17 *Specification for carbon steel electrodes and fluxes for submerged-arc welding*, on the basis of chemical composition but the flux is defined by the weld metal properties which may be obtained from a given wire/flux combination. This means that the same flux can have a different classification depending on the wire with which it is used.

In the German specification, DIN 32 522 *Fluxes for submerged-arc welding: designation, technical terms of delivery*, the fluxes are classified according to their method of manufacture, mineral constituents, type of application, loss or gain of alloying elements, type of current and maximum current and any special characteristics.

The UK specification, BS 4165 *Electrode wires and fluxes for the submerged-arc welding of carbon steel and medium tensile steel* specifies requirements for solid wires and wire/flux combinations to weld steels with tensile strengths up to 700 N/mm^2, using two run or multirun techniques. An appendix describes flux types and characteristics, especially basicity and the wire/flux combinations are graded as complying with one of three strength levels.

Studies of weld metal properties were accompanied by examinations of heat affected zones when it was found that wire and flux combinations capable of producing high quality weld metals had to be used sometimes with restrictions on deposition rate if the HAZ mechanical property requirements for the steel were to be met. A further factor which almost certainly restricted the potential deposition rate of the process was the development and refinement of NDT techniques such as ultrasonic testing of welds and plate materials. Defects that previously escaped detection were now picked out easily and usually had to be repaired. Another troublesome problem was the tendency to large grain size which made ultrasonic inspection difficult because it interfered with beam transmission.

Factors affecting hot and cold cracking were studied in depth and the influence of the welding process, electrode wires and parent material were more easily understood. The destructive influence of hydrogen in weld metal and the HAZ led to extended research work into flux and wire production and measuring techniques for the detection of weld metal and potential hydrogen content in welding consumables.

A good example of the severe steel and weld metal requirements successfully met over this period was the fabrication of offshore oil platforms where the node area requirements in particular specified the highest level of steel and weld metal metallurgical properties.

Applications

The progress that has been made in use of the submerged-arc process is indicated by examining some of the changes which have occurred in submerged-arc welding applications mentioned previously:

1 Pipe welding
Three, four or five wire techniques employing one or two passes are now widely used for longitudinal welds in pipe. There are now specifications covering chemical analysis and mechanical properties for parent materials, weld metals, and HAZs. These include low temperature Charpy V impact test requirements. The extended successful use of submerged-arc in this application was possible only because of improvements to steels and consumables, in particular, the ability of the flux to meet both the high process speed and the metallurgical requirements.

2 Shipbuilding
The most notable change has been the introduction of higher tensile steels, however, there is a more general use of smaller welds and equipment has become more compact and manoeuvrable, Fig.1.3. To meet process and mechanical properties requirements, including Charpy V impact values, new fluxes were developed enabling high current welding procedures to be maintained. By the 1970s multipower welding with three and even four wires was an established procedure for panel lines.

16

1.3 *Submerged-arc head on a tractor with flux recovery unit.*

3 Bridge building

The development of wire and flux combinations depositing weld metal with a good notch toughness has made it possible to extend the use of this process on butt welding applications previously possible only with basic MMA electrodes and the now obsolete 'Fusarc' continuous electrode welding process.

4 Pressure vessel construction

In this area, which includes fabrications such as jacket sections of offshore platforms, the high heat input techniques used in the 1940s were severely restricted because of the much more stringent weld metal and HAZ requirements. Development work has improved deposition rates while still maintaining metallurgical requirements.

5 Rebuilding and surfacing

Parent metal and weld metal metallurgy has received considerable attention and knowledge of the metallurgical changes occurring during surfacing has enabled the materials to be more accurately classified. Use of strip and hot wire welding has increased deposition rates appreciably. Parallel welding, series arc, DCEN, long stickout and metal powder additions are now everyday techniques.

17

6 Motor industry

The major changes have centred on the use of more robotics which has led to many submerged-arc applications being taken over by MIG welding.

7 General fabrication

Some of the wire and flux combinations available in 1940 are still in use today, *e.g.* the manganese-silicate type of flux which, with 0.5%Mn welding wire is still one of the cheapest wire and flux combinations for welding ordinary mild steel. Thicknesses welded range from 3mm upward.

CHAPTER 2
Principles

The principle of the submerged-arc process has been shown in Fig.1.2. A power source is connected across the contact nozzle on the welding head and the workpiece. The power source can be a transformer for AC welding or a rectifier (or motor generator) for DC welding. The consumables are a bare, possibly coppered, continuous electrode and a granular welding flux fed to the joint through a hose from a flux hopper. To prevent the electrode overheating at high currents the welding current is transferred to the electrode at a point close to the arc. The arc burns in a cavity filled with gas (CO_2, CO, etc) formed by breakdown of the flux, and metal fumes. The cavity is contained ahead of the arc by unfused parent material and behind by the solidifying weld metal. The covering over the cavity consists of molten slag. The figure also shows the solidified weld and the thin covering of solid slag which has to be detached after the completion of each run. Because the arc is completely submerged by flux there is none of the irritating arc radiation which is so characteristic of the open arc processes like MMA and MIG and welding screens are therefore unnecessary. There is an almost complete absence of smoke and fume. The welding flux is never completely consumed and that which is unfused can be collected, either by hand or automatically, and returned to the flux hopper to be used again.

Although semi-automatic submerged-arc welding equipment exists and is convenient for certain applications, most of the submerged-arc welding carried out today makes use of fully mechanised welding equipment. Indeed one of the main virtues of the submerged-arc process is the ease with which it can be incorporated into fully mechanised welding systems to give high deposition rates and consistent weld quality. Weld metal recovery approaches 100% as losses through spatter are extremely small. Heat losses from the arc are also quite low because of the insulating effect of the flux cover, and for this reason the thermal efficiency of submerged-arc can be as high as 60% compared with about 25% for MMA welding. (Account must be taken of this when applying heat input calculations derived for MMA to submerged-arc, see section on hydrogen cracking later.) The weight of flux fused during welding is approximately equal to the weight of wire consumed, the actual ratio — weight of wire fused to weight of flux fused — being dependent on flux type and welding parameters used. Not all the flux laid down ahead of the weld is fused and that unfused can be recycled so the actual weight of flux used during welding depends on the efficiency of the flux recovery methods.

Equipment

Wire feeders and power sources

A feedback system is generally used to maintain a stable arc length. A change in arc length causes a corresponding change in arc voltage which is used to produce an increase or decrease in the wire feed rate until the original arc length is regained. This method requires a constant current (drooping) power source. If small diameter wires are used wire feed rates are higher and it is then possible to employ a constant wire speed and to make use of the self-adjusting effect of a power source to provide an inherent control of arc length. With a constant wire feed rate a constant voltage (flat) power source is required. A steady arc length is maintained because a constant voltage power source gives appreciable changes in current with changes in arc voltage and these alter the burnoff rate in such a way as to oppose the changes in arc length. The welding current is then increased by increasing the wire feed rate or decreased by decreasing the feed rate while keeping the power source voltage constant.

Over the years control circuitry used in welding machines has taken advantage of the latest developments in electronics, with the result that welding heads and control boxes are now more compact and versatile. Some submerged-arc machines are supplied as modules which means that the user can purchase separately the wire feed motors, control boxes, flux hoppers and accessories necessary to make up a welding unit suitable for a particular type of work. In some units the feedback control can be switched out so that fine diameter wires can be fed at relatively high speeds as may be needed for high speed welding of thin material. This facility to switch out the arc length control unit is useful when the welding head may also be required for fully automatic MIG welding with a water cooled gun in place of the submerged-arc feed tube and contact tip. A wire diameter not greater than 2mm would be required and as the wire feed rate could then be quite high the particular feeder, if it is not part of a purpose built unit, must have the necessary speed range. In submerged-arc welding with wire diameters of 2.4mm and above, the self-adjusting action operates too sluggishly for the constant wire speed/constant voltage power source system to be used. Figure 2.1 shows a typical tractor mounted electrode feed unit and control box which governs arc voltage and welding current. Figure 2.2 is a block diagram of a control circuit. The main part of the arc voltage control is identified as the regulation circuit. A simplified description of its function is as follows — a thyristor regulator feeds the armature of the motor with a voltage directly proportional to the setting on a potentiometer and in addition keeps the set motor speed constant irrespective of loading. This is achieved by comparing the reference voltage obtained from the potentiometer with two voltages fed back from the armature circuit of the motor. One of those is the armature current voltage drop across a resistance and the other corresponds to the induced voltage in the armature which is directly proportional to the motor speed.

If the speed of the motor changes, *e.g.* because of a change in load, a voltage difference is produced in the comparison circuit. This voltage is amplified and influences the ignition point of the thyristors via a pulse unit, so that the voltage output from the thyristors corrects the speed

2.1 *SAW equipment on a self-steering tractor showing flux hopper and process control box.*

21

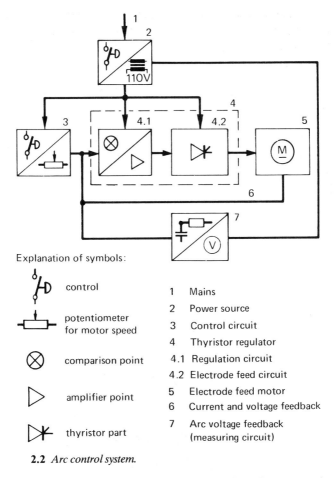

Explanation of symbols:

control

potentiometer
for motor speed

comparison point

amplifier point

thyristor part

1   Mains
2   Power source
3   Control circuit
4   Thyristor regulator
4.1 Regulation circuit
4.2 Electrode feed circuit
5   Electrode feed motor
6   Current and voltage feedback
7   Arc voltage feedback
    (measuring circuit)

2.2 *Arc control system.*

change. In addition the arc voltage is fed back to the comparison circuit giving a constant arc length between electrode tip and workpiece.

When the distance changes between electrode tip and workpiece the arc voltage also changes, which results in a voltage difference in the comparison circuit. This voltage difference causes a change in motor speed which corrects the distance between the electrode tip and the workpiece. The feedback of the arc voltage also facilitates arc striking when starting to weld but this feedback comes into play only when the thyristor regulator is used for electrode feed.

Many welding heads also have facilities for feeding the wire fast irrespective of the position of the potentiometer. This is useful when setting up the machine for welding or when a new coil of wire has been fitted. For final positioning of the wire in the joint there is an 'inching' facility which permits the wire to be approached to the joint in a slow controlled manner.

The current is fed to the electrode wire about 25mm from the striking end. This means that, compared with an MMA electrode, much higher welding currents can be used without the electrode overheating, *e.g.* the maximum welding current which can be used on a mild steel 4mm submerged-arc wire is about 800A compared with about 190A for a 4mm MMA electrode. Both penetration and deposition rate are thus higher with the submerged-arc process than MMA.

2.3 *Wire feeder and flux recovery unit in which the flux is cleaned of fused flux, dust and fines.*

Accessories

Almost all welding heads have a wire straightening device *e.g.* a group of three rollers, just in front of the wire feed rolls to facilitate smooth feeding of the electrode. A simple indicator, a lamp (as in Fig.1.3) or pointer, is fixed to the guide tube so that the position of the electrode wire under the flux may be judged.

A flux recovery unit is a useful accessory not only to economise on flux but also to limit the spread of flux into the bearings of positioners and electric motors and sometimes also the electrode contact tip where it causes stray arcing and overheating. A suction nozzle of suitable shape is positioned about 100mm behind the arc and as the welding head traverses the joint the unfused flux is sucked up, cleaned and returned to the flux hopper, Fig.2.3. For large installations, panel lines and roll

2.4 *Welding head with motorised cross slides and probe for electromechanical guidance system.*

24

cladding units, special high capacity flux recovery systems, remote from the welding head, are available which can handle and store large volumes of hot flux, see Chapter 4.

Other useful accessories are cross slides on which the welding head can be mounted thus enabling the welding head to be moved in both horizontal and vertical directions. This greatly simplifies setting up for welding. The slides may be manually operated or motorised, the latter sometimes being used in conjunction with a seam tracking unit which keeps the electrode correctly positioned in the joint, see Fig.2.4.

Joint preparation and welding procedure

Before starting production welding it is essential to be properly prepared if costs are to be kept as low as possible and production rates maximised. This involves making decisions regarding joint preparation, joint fit-up, welding procedure, selection of welding data and inspection.

Joint preparation depends on plate thickness, type of joint, *e.g.* circumferential or longitudinal, and to some extent on the standards to which the structure is being made. Plates up to 14mm thickness can be butt welded without preparation using one or two sided welding with a gap not exceeding 1mm or 10% of plate thickness, whichever is smaller. An edge preparation may be needed if a build-up of reinforcement must be avoided. Thicker plates also need preparation if full penetration is to be obtained. As with any automatic welding process variable fit-up cannot be tolerated. With MMA the welder's technique can be adjusted to cope with variations in fit-up and root face dimension but this is not possible with an automatic process like submerged-arc. If conditions are set up for a root gap of 0.5mm and this increases to 2-3mm, burnthrough will occur unless an efficient backing strip is used. If variations in root gap cannot be avoided a manual root run using MIG or MMA electrodes is advisable. All plate edges must be clean and free from rust, oil, millscale, paint, *etc.* In fillet welds the faying surfaces should also be clean as they are partly fused during welding. If impurities are present and are melted into the weld, porosity and cracking can occur. Time spent in minimising such defects by careful joint preparation, accurate setting-up and thorough inspection before welding is time well spent as weld defects are expensive to rectify and cause delays.

The responsibility for welding procedure which includes weld sequence is usually a matter for the designer and welding engineer who should know whether or not preheat is necessary and if a single pass or multipass weld is required. In general, the more severe the requirements for low temperature notch toughness the lower the maximum heat input and therefore welding current that can be used. This may mean that a multipass technique is necessary. When welding stainless steels heat input should be kept low for other reasons: stainless steel has poor thermal conductivity and a high coefficient of expansion compared with mild steel. These two effects lead to overheating and excessive distortion if large diameter wires and high currents are used. Multirun welds using small diameter wires are therefore recommended for stainless steels and high nickel alloys such as Inconel. Stainless and high alloy wires have high electrical resistivity and can become overheated with high currents or long electrode extensions.

Welding conditions

Selection of correct welding conditions for the plate thickness and joint preparation to be welded is important if satisfactory joints free from defects such as cracking, porosity and undercut are to be obtained. Welding conditions also affect bead size, bead shape, depth of penetration and sometimes the chemical composition of the deposited metal.

Electrode polarity

The deepest penetration is usually obtained with the DCEP polarity which also gives the best surface appearance, bead shape and resistance to porosity. Direct current electrode negative polarity gives faster burnoff and decreased penetration as maximum heat is developed at the tip of the electrode instead of at the surface of the plate. For this reason DCEN polarity is often used when welding steels of limited weldability and when surfacing as, in both instances, penetration into the parent material must be kept as low as possible. The flux/wire consumption ratio is less with DCEN polarity than with DCEP so transfer of alloying elements from the flux is reduced.

In changing from DCEP to DCEN polarity some increase in arc voltage may be necessary to obtain a comparable bead shape. Alternating current requires a voltage about halfway between DCEP and DCEN polarity and is particularly useful when arc blow is a problem. It is often used in tandem arc systems in which DCEN is used on the leading electrode and AC electrode on the trailing electrode (see also Chapter 3).

Welding current

The deposition rate increases as welding current increases and current density determines depth of penetration, the higher the current density the greater the penetration for the same current. For a given flux, arc stability is lost below a minimum current density so if the current for a given electrode diameter is too low a ragged irregular bead is obtained. Too high a current density also leads to instability because the electrode overheats. Undercutting may also occur.

Figure 2.5 shows the welding current range for mild steel wires of diameter 1.6-6.0mm, together with deposition rates in kg/hr at the maximum and minimum welding currents using DCEP polarity. On DCEN polarity deposition rates are approximately 35% greater at any particular current. Welding current should be selected to give the penetration required. The deposition rates shown apply to mild steel and low alloy steel electrodes, but not to stainless steels, high nickel wires or non-ferrous materials.

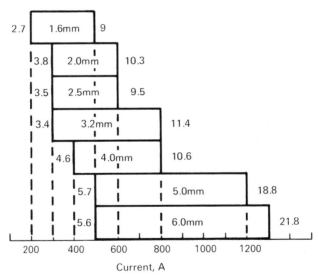

2.5 *Recommended (DCEP) current range and deposition rates (kg/hr) for different wire diameters.*

Electrode diameter

The commonly used submerged-arc wire diameters lie in the range 2.0-6.0mm. Current ranges are wide for solid wires, as shown in Fig.2.5, and the overlap in these ranges enables exploitation of wire diameter

27

effects, *i.e.* at the same current, a smaller diameter wire gives a deeper penetrating and narrower weld bead than a larger diameter wire because of the current concentration effect. Arc starting and stability are also generally superior with wire of smaller diameter. Cored wires are finding increasing application for submerged-arc welding and are usually at the lower end of the size scale.

Arc voltage

Arc voltage has an important effect on weld bead shape, raising the voltage producing a wider flatter bead. The effect of arc voltage is often misunderstood because it affects dilution rather than penetration. Bead-on-plate welds and square edge close butt welds (no gap) have increased width and dilution as arc voltage increases, but depth of penetration is less affected, Fig.2.6 *a*. If the joint is 'open', as for example in

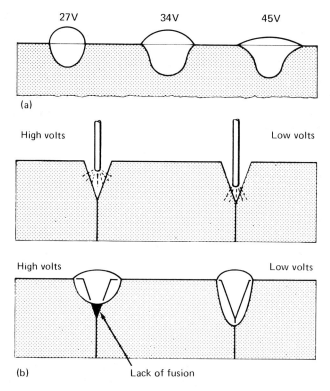

2.6 *Effect of arc voltage on bead shape:* a) *Bead-on-plate and square edge close butt welds;* b) *Prepared joints.*

a butt joint with a small angle V preparation, arc voltage variations can affect penetration at the same current, Fig.2.6 b.

Increasing arc voltage lengthens the arc so that weld bead width, reinforcement and flux consumption are increased as is the probability of arc blow. When alloying the weld metal from the flux, arc length and hence arc voltage must be carefully controlled as at high arc voltages more flux is melted allowing more alloying elements to enter the weld metal thereby affecting weld metal composition.

Welding speed

Bead size is inversely proportional to welding speed at the same current. Higher speeds reduce bead width, increase the likelihood of porosity, and if taken to the extreme, produce undercutting and irregular beads. At high welding speeds the arc voltage should be kept low otherwise arc blow is likely to occur. If welding speed is too low burnthrough can occur. A combination of high arc voltage and low welding speed can produce a mushroom-shaped weld bead with solidification cracks at the bead sides, Fig.2.7. For a given arrangement of wires and wire diameters welding speed is limited by the welding current which can be tolerated by the flux. Some fluxes are specially compounded to allow high speed operation. Higher speeds are possible with twin wire operation or by holding a more acute electrode angle.

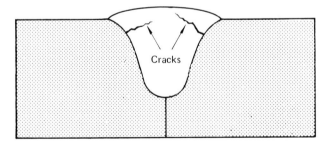

2.7 *Mushroom shaped weld penetration resulting from high voltage combined with low speed.*

Multiple wire techniques are used for both butt and fillet welding and there is a marked speed advantage with an increase in the number of wires. This is illustrated in Fig.3.20. It is necessary to change welding conditions as thickness increases, with heat input being in the ratio 1:2:3 for one, two and three wires respectively.

Electrode extension (stickout)

Electrode extension is an important variable as it governs the amount of resistance heating which occurs in the electrode. If the extension is short the heating effect is small and penetration is deep. Increasing the

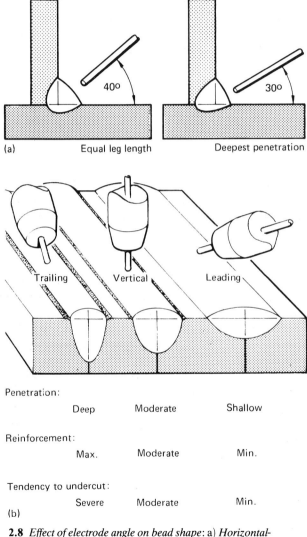

(a) Equal leg length   Deepest penetration

Penetration:
Deep   Moderate   Shallow

Reinforcement:
Max.   Moderate   Min.

Tendency to undercut:
Severe   Moderate   Min.

(b)

2.8 *Effect of electrode angle on bead shape*: a) *Horizontal-vertical fillet welds*; b) *Comparison of trailing, vertical and leading arcs.*

extension increases the temperature of the electrode but decreases the penetration although deposition rate is increased. Increased extension is therefore useful in cladding and surfacing applications but steps have to be taken to guide the extended electrode otherwise the arc may wander. For normal welding, electrode extension should be 25-30mm for mild steel and rather less, say 20-25mm, for stainless steel. This is because the electrical resistivity of stainless wire is appreciably greater than that of mild steel wire. Cored wires also have a higher resistivity because some of the cross section is occupied by non-conducting flux.

Electrode angle

As the angle between electrode and plate determines the point of application and direction of the arc force and hence weld pool motion, it has a profound effect on both penetration and undercut. Figure 2.8*a* shows this effect on horizontal/vertical fillet welds, and Fig.2.8*b* compares the effect obtained with a vertical arc with those obtained with leading and trailing arcs. The effect on undercutting can be particularly marked and a more acute angle allows higher speeds to be used.

Flux depth

The depth of the flux is often ignored and the powder is simply heaped around the wire until the arc is completely covered. For optimum results the flux depth should be just sufficient to cover the arc although, at the point where the electrode enters the flux cover, light reflected from the arc should be just visible. With too shallow a flux cover the arc may flash through and can cause porosity and a rough surface because of inadequate metallurgical protection of the molten metal. Too deep a flux cover gives a worse bead appearance and can lead to spillage on circumferential welds. On deep preparations in thick plate it is particularly important to avoid an excessive flux cover as weld bead shape and slag removal can be unsatisfactory.

Special techniques

Circumferential seams

When welding circumferential seams on pipes or pressure vessels, or when surfacing cylindrical objects, care must be taken to prevent spillage

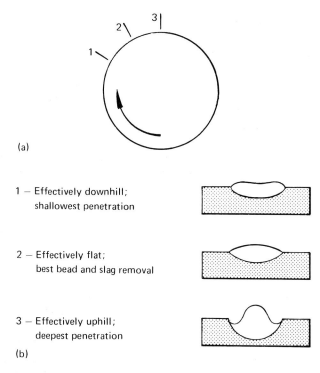

(a)

1 — Effectively downhill;
   shallowest penetration

2 — Effectively flat;
   best bead and slag removal

3 — Effectively uphill;
   deepest penetration

(b)

2.9 *Electrode positions for circumferential welding and resulting bead shapes.*

Table 2.1 Electrode displacements from TDC for circumferential welding

External diameter of workpiece, mm	Electrode displacement, mm
25-75	12
75-450	22
450-900	34
900-1050	40
1050-1200	50
1200-1800	55
> 1800	75

of the molten pool and loss of flux covering. Some form of support for the flux may therefore be necessary.

Figure 2.9 shows three possible positions of the electrode and of these, position 2 usually gives the best results as regards bead appearance and slag removal. Position 1 gives the worst bead appearance and such an arrangement can be difficult to control on small diameter workpieces as the molten slag tends to run ahead of the arc.

Table 2.1 gives recommended electrode displacements relative to the vertical position for various workpiece diameters. For any particular application, displacement depends upon wall thickness (or mass in a solid object), welding current and welding speed but values given serve as a guide.

Strip cladding

Although most applications of submerged-arc involve single or multiwire systems using round wires, electrodes in the form of strip are often used for cladding. Strips are usually 0.5mm thickness, the most common strip width being 60mm but wider strips, *e.g.* 100mm, can be used without any loss of quality. Strip cladding has the advantage that penetration is low particularly with DCEN and deposition rate is relatively high. Modern fluxes designed for strip cladding have greater current tolerance than earlier types and use of currents up to 1200A with austenitic stainless steel strips gives deposition rates of up to 22 kg/hr with DCEP and 32 kg/hr with DCEN. Inconel can also be deposited and provided the flux is a low silica type and the Inconel strip used contains 2-3%Nb good quality crack-free deposits can be obtained. Monel, aluminium bronze, nickel and 13%Cr strips have also been successfully used as strip cladding electrodes. Good electrical contact between the strip and contact tip is essential.

Weld defects

Weld defects, such as poor weld bead shape, lack of fusion, slag inclusions and porosity are normally avoided by adherence to the correct welding procedures. In multirun welds careful placement of beads is necessary to avoid creation of deep narrow gaps between the weld beads and the fusion faces in which slag might become trapped. A few defects are not directly related to welding conditions and, with these, good housekeeping particularly of the flux, wire and weld preparation also plays its part.

33

Porosity

Porosity is a fairly common defect which can be influenced by many factors. Sometimes it is clearly visible as pinholes in the weld surface, at other times it is below the surface and is revealed only by X-ray examination or ultrasonic testing. Unless it is gross or preferentially aligned, porosity is not likely to be harmful.

Common causes of porosity are:

1 Contamination of joint surfaces with oil, paint, grease, hydrated oxides, *etc.* These decompose in the arc to give gaseous products which can cause elongated 'wormhole' porosity often located along the centreline of the weld.

2 Damp flux: flux should be kept dry. It is good practice to dry all fluxes before use and store them in a heated building. The manufacturer's recommendations regarding drying temperatures should be observed. Note that if a flux recovery unit, driven by compressed air, is used the compressed air should be dried thoroughly.

The surface of a weld may sometimes contain small depressions known as surface pocking or gas flats. These are harmless and while the exact cause is not fully understood it has to do with conditions which cause generation of gas or make it difficult for gas to escape, for example, moisture or lack of deoxidants and too many fines in the flux to allow gas to pass readily.

Hydrogen-induced cracking

Some steels are more susceptible to hydrogen-induced cracking than others but fortunately submerged-arc welds are not particularly prone to this type of defect. If the steel has a high hardenability, with a carbon equivalent of over 0.40, however, and the flux is damp, cracking can occur. The carbon equivalent is calculated from one of several formulae which take account of the hardening effect of alloying elements. A widely used formula is that of the International Institute of Welding as follows:

$$CE = C\% + \frac{Mn\%}{6} + \frac{(Cr\% + Mo\% + V\%)}{5} + \frac{(Ni\% + Cu\%)}{15}$$

Adherence to recommendations for handling flux from consumable manufacturers and to preheat, interpass or post-heat procedures should enable hydrogen cracking to be avoided, see Chapter 4.

Important factors in avoidance of hydrogen cracking are use of clean wire and dry flux to minimise hydrogen input to the weld. In addition it may be necessary to control the size of the weld bead by maintaining a minimum energy input to avoid a hardened HAZ. Where this is not possible preheat may be needed to allow sufficient hydrogen to escape from the weld to avoid cracking. The submerged-arc process is thermally very efficient and the standard BS 5135 *Metal arc welding of carbon manganese steels* takes account of this when SAW is used by introducing a factor to the preheat and heat input tables for avoidance of hydrogen cracking produced initially for MMA welding.

On the whole, submerged-arc welds are less liable to suffer hydrogen cracking than MMA welds, probably because the energy input is usually larger. However, it is not possible to achieve such low weld hydrogen content as with MMA electrodes dried at high temperatures or with gas shielded welding. When hydrogen cracking does occur in submerged-arc welds it is frequently in the weld metal rather than the HAZ. These weld metal cracks can be difficult to detect if present as transverse cracks at 45° to the weld surface unless special ultrasonic testing techniques are used, see Fig.2.10. Cracking of this type is sometimes called 'chevron cracking' and it was extensively investigated in the 1970s. It was shown that weld metal hydrogen resulting from flux moisture was a major factor. The defect occurred readily with the higher basicity agglomerated fluxes but with sufficient moisture present it could also occur with lower basicity and/or fused fluxes.

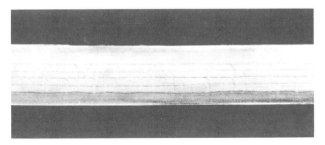

2.10 45° (*chevron*) *cracking.*

Solidification cracking

Because of the large weld pools and high welding speeds used with SAW, hot cracking may be encountered and is usually found along the centreline of the weld. When it occurs the penetration profile should be

35

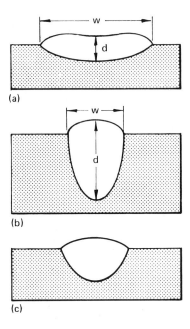

2.11 *Form factor for SA weld beads:* a) *W* > *d giving tendency for surface cracks;* b) *W* < *d giving tendency for centreline cracking;* c) *W/d* ≈ 3/2 *giving sound welds.*

examined. Deep narrow welds are prone to centreline cracking, Fig.2.11. Mushroom-shaped beads as shown in Fig.2.7 should also be avoided.

Solidification cracking is controlled by the composition of the weld, its solidification pattern and the strain on the solidifying weld metal. The problem is aggravated by the presence of phosphorus, sulphur and carbon and if these elements are known to be present in the parent material in higher amounts than usual, a change should be made to a wire with a higher manganese content and steps taken to minimise dilution and ensure good weld bead profiles. The most dangerous element is carbon which, if other considerations allow, can be kept low in the weld by use of high silica fluxes, *i.e.* Mn and calcium silicate types. BS 5135 gives a formula for predicting the cracking tendency of weld compositions. Cracking is normally a problem only in the root runs, as in Fig.2.12*b*, where dilution of parent plate into the weld is high giving high carbon contents. Long deep weld pools as in Fig.2.12*a* or welds made at high welding speeds or with high restraint and large root gaps as in Fig.2.12*c*, accentuate the problem.

36

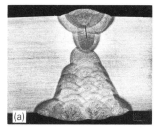

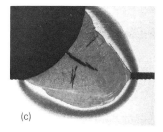

2.12 *Solidification cracking*:
a) *In the root beads of a multi-run weld*; b) *Caused by high speed giving a long deep weld pool in first pass*; c) *Caused by high restraint and root gap.*

Occasionally a groove may be found on the surface running along the centre of the weld. This may be caused by shrinkage and although it is sometimes mistaken for incipient solidification cracking it is actually only superficial.

Arc blow

Arc blow is caused by magnetic forces around the arc which deflect it causing defective welds. It occurs only with DC. Although the electrode may be correctly positioned in the joint the arc may wander and the deposited weld become misshapen. A good and correctly positioned earth is essential and the problem mentioned above can usually be cured by moving the earthing point and/or using a second earth.

The topics which have been considered in this chapter are common to all submerged-arc welding applications and to the variants of the process which will be considered in more detail in the chapters which follow.

CHAPTER 3
Process variants

The earlier chapters indicated that the submerged-arc process has undergone considerable development and that many different ways of using the basic process are now available. With the basic process single wires from 2.0-6.0mm diameter may be used with either DC or AC power. Smaller diameter wires are sometimes employed when welding sheet metal and, although with this thickness MIG processes may be preferred, submerged-arc does have advantages such as absence of spatter and no light or fume from the arc. The process is almost always used in the automatic form but a manual version also exists. In addition to these variants the single electrode process has also been used with different electrode shapes and polarities, long electrode extensions and the addition of extra metal. This can be provided by cold and hot wire techniques or through comminuted or powdered metal additions. Other important variants use more than one electrode. Narrow gap welding techniques are also available.

Single electrode welding

Fine wire welding

Equipment for submerged-arc welding has evolved to provide a versatile range of machines. Machines for use with thick wire rely on arc voltage to control the feed rate to give arc length control at virtually constant current. For smaller wire diameters this system has been largely replaced by constant wire feed rate in combination with constant voltage power sources, as discussed in Chapter 2. In these machines, arc length control is derived from rapid changes in welding current because of the self-adjustment effect inherent in the power source. Constant voltage systems are superior at the high wire feed rates required with small diameter wires. Multiprocess machines are now available with a choice of response characteristics suitable for fine or larger wire submerged-arc welding as well as flux-cored wire and MIG welding.

The development of fine wire submerged-arc welding with DCEP using one or two wires allows the process to be applied to a wide range of light fabrications. Approximate stickout length, current, current density, and voltage for fine wire SAW are given in Table 3.1 and these figures are similar to, but have a slightly wider range than, figures for the MIG spray arc.

Table 3.1 Welding conditions for small diameter wires

Wire diameter, mm	Stickout, mm	Current, A	Current density, A/mm	Voltage, V
0.8	12	100-200	200-400	20-25
1.2	20	150-300	130-265	22-30
1.6	20	200-500	100-250	23-36
2.0	25	200-600	65-190	30-40

Figure 3.1 shows lap welding of a bung into an LPG cylinder using DCEP with a single 2.4mm wire at 300A. Figure 3.2 is of butt welding of the two halves of the same LPG cylinder using tandem parallel arc DCEP with two 1.2mm wires at 600A. Submerged-arc welding is useful for welding sheet because it is capable of high welding speeds that minimise distortion and the weld bead is smooth and free from spatter. Productivity in sheet welding depends on welding speed, not deposition rate, and fluxes are available for welding speeds up to about 3750 mm/min with a single arc and 5000 mm/min with twin arcs. These speeds can be achieved only with precise equipment and accurate joint location. The best results are obtained by using equipment suited for the required current. For example, when making welds at about 200A it is better to use a good heavy duty MIG power source rather than a 1000A submerged-arc machine. Care should be taken, however, not to use machines at currents above their 100% duty cycle rating because submerged-arc welding frequently runs continuously for over 10min (the time on which ratings are often based).

Stainless steel sheet welding is a particularly good application for fine wire SAW because weld finish is spatter free and often good enough to eliminate the need for grinding and polishing with significant cost benefits.

3.1 *Machine for welding bungs into LPG cylinders. (Lap weld, 300 A (DCEP), 28 V, 32 mm stickout, 2.4 mm wire, welding time 20 sec.)*

Semi-automatic submerged-arc welding

In semi-automatic SAW the gun is similar to that used in gas shielded welding, except that flux is fed to the gun to shield the arc, instead of gas. The wire and flux are fed through the handle of the gun and the flux falls through a conical nozzle to surround the arc. Fluxes for semi-automatic welding require a higher proportion of fine particles than for automatic welding. Such fluxes can be propelled, by air pressure, from a tank through a hose which can be as long as 20m. If it is not convenient to feed flux from a pressurised tank, or if the flux is not formulated to flow readily, a small hopper, holding about 1.3kg of flux can be mounted directly on the torch.

The semi-automatic submerged-arc gun can be operated manually, as in Fig.3.3, or clamped to a tractor or simple fixture. The arc and weld pool

40

3.2 *Machine for butt welding LPG cylinders. (Joggled butt 2.4mm thickness, 590A (DCEP), 27V, 25-38mm stickout, twin 1.2mm wires, parallel, time 42sec.)*

cannot be seen but for fillet welds the arc can be placed accurately by pressing the flux cone into the angle of the fillet and dragging the torch at the required speed. In butt welds the torch can be guided by eye, which is not difficult or critical because the weld pool is comparatively large. A supplementary guide, such as a piece of angled steel clamped to the workpiece can be used if necessary. Holding the correct travel speed is not difficult for a skilled welder but improved control can be obtained with a small motor attached to the gun which drives a knurled wheel that runs along the surface. Alternatively the gun may be mounted on a small self-propelled tractor.

3.3 *Semi-automatic SAW.*

41

Semi-automatic submerged-arc welding is a highly productive process capable of deposition rates as high as 10 kg/hr. No other arc welding process is as comfortable to use manually with currents as high as 500 A as there is no fume or flash and little heat.

Electrode polarity

Direct current can be supplied from a motor generator but increasingly transformer-rectifiers are used. The electrode positive polarity gives good penetration and reproducibility. Alternating current gives less penetration than DCEP but its principal advantage is its freedom from arc blow.

If the electrode is made negative (DCEN) deposition rate is increased significantly, see Fig.3.4, because of the cathodic heating of the wire. This increase in deposition rate is achieved without additional capital expenditure and fluxes suitable for DCEP are generally suitable also for DCEN. Power sources having variable voltage characteristics or a constant voltage with inductance may be used but experience has indicated that constant voltage power sources without inductance are not satisfactory because of arc instability. An arc voltage about two volts higher for 4mm diameter wire than that usually used for DCEP at the same current is necessary.

There is a decrease in penetration of between 20 and 25% when using DCEN compared with that normally achieved by DCEP. This is not usually significant except when specific root penetration is required.

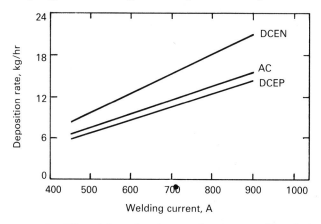

3.4 *Effect of electrode polarity on deposition rate. (Standard electrode extension, 32 mm, 4 mm diameter wire, 32 V.)*

42

When fillet welding, North American and European welding specifications permit a reduction of fillet weld leg length from that required by design, provided that the throat thickness required by design can be achieved. This means root penetration is crucial and in these circumstances DCEN should not be used.

In multipass welding placing of individual weld beads is more critical with DCEN than with DCEP because the reduced penetration can result in lack of fusion or interpass slag traps. For fillet welding the lower penetration is usually not significant, unless use is being made of the permitted reduction in leg length mentioned above.

The increased bulk of deposited metal resulting from the higher deposition rate with DCEN should be accommodated by increasing the welding speed rather than by allowing the bead to become thicker. Because bead placement is most important the increased speed can also result in tracking and bead alignment problems. Maintaining adequate preheat, where required, at the higher travel speeds can be more difficult. Finally, retraining of operators is essential to avoid defects resulting from the lower penetration and higher welding speeds.

Long stickout welding

Wire burnoff rates can be raised by increasing the electrical extension (the amount of wire protruding from the contact tip towards the work) from the usual 30mm to some greater length, *e.g.* 165mm, as shown in Fig.3.5.

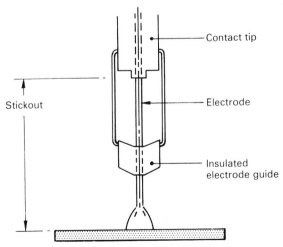

3.5 *Long stickout welding (LSO).*

This extension is usually called stickout and for convenience it is measured from the contact tip to the work although to be strictly accurate it should be the distance from the contact tip to the root of the arc. It is not a bad approximation, however, because the arc is usually buried beneath the plate surface.

The wire protruding beyond the contact tip is heated according to the relationship:

$$\text{Heat}\,(H) = \frac{I^2 L \rho}{D}$$

where I is the current
       L is the length of stickout
       ρ is resistivity of the wire (varies with temperature)
       D is diameter of the wire

The degree of preheating of the wire with its associated increase in burnoff rate therefore increases with increasing current and length of stickout, and decreasing wire diameter. Deposition data for various contact tip/plate distances are shown in Fig.3.6, while Fig.3.7 is a comparison of the deposition rates for traditional submerged-arc welding with DCEP with submerged-arc welding utilising both DCEN and long stickout (LSO) techniques.

In conventional submerged-arc welding the 'indicated arc voltage' shown on the welding machine meter is the voltage drop from the contact tip to

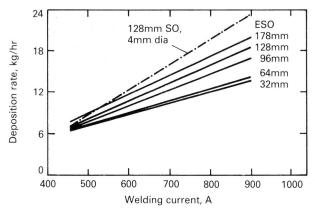

3.6 *Effect of stickout on deposition rate, (DCEP), 4.75 mm diameter wire.*

44

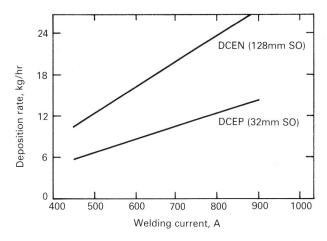

3.7 *Comparison of deposition rates for DCEP and normal stickout with DCEN and long stickout.*

the workpiece, which includes, apart from the arc voltage, the voltage drop caused by contact resistance and that by the resistance of the length of wire beyond the tip. This indicated voltage has been taken by common consent to be representative of the actual arc voltage and is the parameter generally used in setting up welding procedures as it can be conveniently measured. If the length of wire beyond the contact tip is increased the 'indicated arc voltage' increases. Therefore, when setting up welding parameters for LSO welding an increase in indicated arc voltage is necessary to compensate for the voltage drop in the increased length of wire beyond the contact tip. The increase in indicated arc voltage depends on both wire diameter and the extra stickout beyond the standard 32mm. This is illustrated in Table 3.2 for 4mm diameter wire with DC welding currents of between 500 and 1000A.

Table 3.2 Increase in indicated arc voltage for different electrode extensions, 4mm wire

Electrode extension (stickout), mm	Indicated arc voltage
32	V
96	V + 3
128	V + 5

45

With a long stickout the wire becomes soft from resistance heating and the end tends to wander either under the magnetic force of the arc or the natural cast of the wire. As exact alignment of the arc is important in submerged-arc welding, it is necessary to provide an insulated guide for the hot wire to about 25mm above the arc. The limit on the contact tip/plate distance which can be used varies with wire diameter as shown in Table 3.3.

Table 3.3 Maximum contact tip/plate distance for different wire diameters

Wire diameter, mm	Electrode extension (stickout), mm
3.2	76
4	128
4.75	165

The practical limits given in Table 3.3 stem from the problems of passing the hot flexible wire through the bottom guide without jamming and bending the contact tip and guide and while these limits have been exceeded experimentally the results cannot be considered reliable. It has been observed that, when the contact tip/plate distance exceeds 130mm, DCEN seems more reliable than DCEP. A successful practice is use of 4mm diameter wire with a maximum of 128mm contact tip/plate distance for currents up to 800A on DCEN.

Long stickout welding is not generally used with AC as the welds appear to be susceptible to slag entrapment in narrow preparations and there is a tendency for arc instability on small fillet welds. Apart from a reduction of 10% in penetration, use of LSO welding does not result in any changes to weld bead geometries compared with conventional submerged-arc and the effect of changes in current, speed and voltage are those expected in normal submerged-arc welding. The submerged-arc process used with LSO seems to be more tolerant to poor joint fit-up than with normal stickout and it is believed that this results from the 'softer' arc.

Arc striking is a little more difficult with LSO than with normal stickout and a 45° cut on the wire with a scratch or hot start technique is generally required. Accurate positioning of weld beads is important to obtain adequate penetration and avoid slag traps but apart from additional care, the same rules used for conventional stickout welding apply to LSO welding. The LSO attachment can be fitted easily to most standard equipment at minimal cost and has been successfully used on single and tandem electrode arrangements. Increased deposition rates are utilised to increase travel speed rather than weld layer thickness, in a similar manner to that recommended for submerged-arc welding with DCEN. Long stickout welding has been used successfully in several workshops but its use has declined mainly because of difficulties with arc starting and the lack of fusion defects already mentioned.

Hot wire welding

The normal submerged-arc process may be operated with the addition of an auxiliary resistively heated wire supplied by a separate feed unit. This gives increased weld metal deposition and the possibility, if precise control of welding conditions can be maintained, of changing weld metal composition by using alloyed or cored hot wire. Hot wire welding is used with the tungsten inert gas (TIG) process and its application to submerged-arc welding is similar and potentially equally widespread. The principle can be applied also to many of the submerged-arc variants, Fig.3.8. Typically a 1.6mm wire is fed into the leading edge of the molten pool and resistance heated by a potential of 8-15V from a separate flat characteristic AC transformer. The technique gives deposition rates in excess of those attainable with DCEN and LSO welding. Deposition rates can be increased by up to 70% for a marginal increase in heat input and mechanical properties are not therefore compromised. Although this variant is attractive in principle it is not now widely used in the pressure vessel industry because of the extra costs incurred for capital and

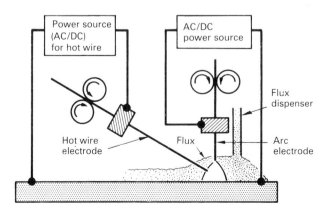

3.8 *Hot wire SAW.*

47

operator training. The additional bulk of the welding head is also sometimes a disadvantage.

Submerged-arc welding with strip electrodes

Instead of using a wire as the consumable electrode a thin wide metal strip (0.5mm thickness, 30-100mm wide) may be employed as in Fig.3.9. Strip electrodes are generally used for surfacing to deposit low penetration, low dilution layers rapidly over a wide area. Most of these techniques therefore use DCEN polarity for a low penetration high deposition arc condition. The arcing point on the strip shifts continually. A dual strip process exists in which an auxiliary cold (not electrically connected) strip is fed into the arc zone of a conventional strip electrode system. This decreases dilution and increases the weld metal deposition rate. Multiple cold wires have been used instead of a single cold strip. As the wires can be cored the deposit composition may be easily varied and this technique is therefore potentially more flexible.

Strip electrodes have also been used for high deposition fillet welding. The strip is turned end on to the direction of travel and not transversely as for surfacing.

Application of single wire methods

Single electrode welding in its various forms is the most widely used type of submerged-arc welding. By selecting the appropriate wire diameter and current type it may be applied to a wide range of single and multipass butt and fillet welds, as well as to surfacing. The composition of wire and flux can be chosen to allow welding of ferrous and non-ferrous metals. Mechanisation of welding when using a single head is effected simply by using a self-propelled tractor, or gantry or by revolving the work under a stationary head. The inclination of the weld seam is limited to $\pm 15°$ because of the danger of flux or molten metal spillage or poor bead shape. Dams to support the flux are used for horizontal welding and for small diameter circumferential welds.

Small lightweight submerged-arc (or multiprocess) tractors have been developed, Fig.3.10, so that high deposition rates and deep penetration can be obtained on welds for which standard equipment is not suited. Even the lightest submerged-arc tractors are capable of welding at up to 1000A and as they weigh less than 16kg can be carried by one man. With such machines, short length butts or H/V fillets can be welded on large fabrications, where the machine must be taken to the work. Excellent

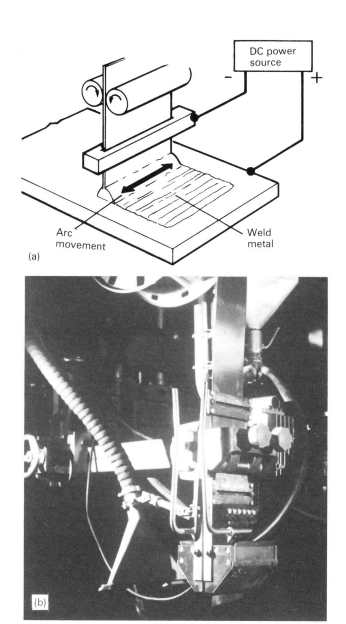

3.9 *Strip cladding*: a) *Principle*; b) *Welding head.*

welds can be made with AC power, simplifying the power source and reducing its cost. A greater tolerance to joint gap variation with AC power

49

3.10 *Lightweight portable tractor.*

follows from the semi-circular penetration profile of an AC weld compared with the finger penetration produced by a DC arc.

The main limitation of the single wire methods is productivity as for each wire diameter there is a maximum welding current and speed above which weld quality deteriorates. This has led to development of the methods described in the following sections.

Multiple electrode welding

Parallel wire welding

In this variant two or more wires are connected in parallel to the same power source, Fig.3.11. When DC is used the arcs converge, DCEP giving the greatest penetration and DCEN the least penetration. With AC the arcs diverge giving medium penetration. Figure 3.12 indicates how electrode alignment affects bead shape — electrodes transverse to welding direction giving shallow penetration and low dilution. With electrodes one behind the other, *i.e.* the tandem arrangement, a higher welding speed is ' possible for a given bead shape than for the corresponding single electrode technique.

The main application for transverse electrodes is surfacing. The electrodes may be given transverse oscillation to minimise dilution and

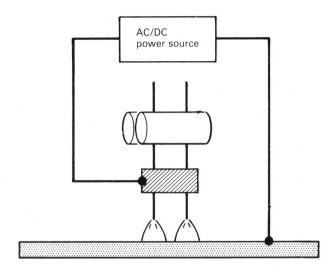

3.11 *Parallel arc welding.*

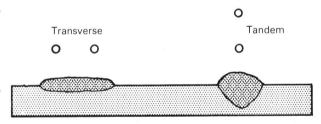

3.12 *Effects of tandem and transverse electrodes on weld penetration.*

reduce heat input and DCEN is commonly used. Tandem parallel electrodes are used for seam welding, as in Fig.3.2, when the welding speed can be approximately 1.5 times that obtainable with a single electrode. Note that the description 'tandem' applies to the electrode alignment with respect to the direction of welding, that is one behind the other. Tandem arrangements are also used with multipower systems when separate power sources are connected to the two wires.

Series arc welding

In series arc, two separately fed electrodes are connected to the opposite poles of a power source, Fig.3.13. Welding arcs are struck between each electrode and molten pool which completes the welding circuit. Direct current arcs have opposite polarity and therefore diverge. The electrode

51

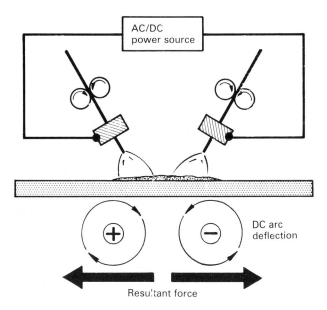

3.13 *Series arc welding.*

positive arc penetrates more deeply than the negative giving an uneven penetration profile. Alternating current gives a penetration bead with constant cross section. When the electrodes are transverse to the welding direction shallow penetration and low dilution are obtained, Fig.3.14. This technique is used for surfacing. Tandem electrode series arc has been used for welding thin material but this is now normally carried out by fine wire SAW or by MIG or cored wire welding. A variation is the shunt wire process in which an earth-connected electrode is introduced into a two or three arc welding system. Each circuit is split between the workpiece and the shunt wire, which melts by direct arc and resistance

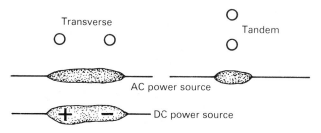

3.14 *Effects of electrode arrangement using series arc.*

52

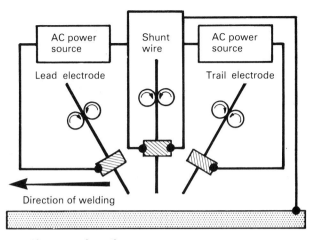

3.15 *Shunt wire electrode arrangement.*

heating. Advantages claimed include increased deposition rate and travel speed together with improved bead appearance, Fig.3.15.

Multipower arc welding

Equipment

Multipower systems use separately powered, driven and controlled electrodes. The processes which make up the final weld bead may, therefore, be distributed among the different arcs, as shown in Fig.3.16.

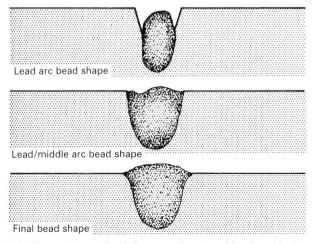

3.16 *Penetration and bead shape as a function of the electrodes in three wire welding.*

53

The lead arc operated at high current and low voltage gives high penetration. The middle arc or arcs operated at lower current than the lead arc adds slightly to penetration and improves bead shape. Finally, the trailing arc or arcs uses lower current and higher voltage than either the lead or middle arcs to smooth and finish the weld bead profile. The lead arc is usually normal to the plate or slightly trailing to optimise penetration, Fig.3.17. A middle arc or arcs is normal to the plate or inclined forward to minimise molten pool disturbance and the trailing arc or arcs is inclined forward to give a smooth weld bead surface. The simplest multipower system widely used for both fillets and butts employs two wires, run in tandem, often with a DCEP lead and an AC trailing electrode.

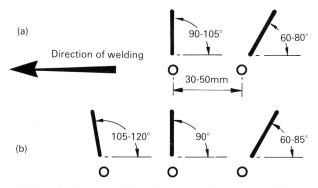

3.17 *Angles for two and three electrode multiarc welding:* a) *Two wire;* b) *Three wire.*

The possible permutations of number of wires, electrode polarity and power source are numerous and the final choice of equipment and welding parameters must be arrived at by careful assessment of all factors. Experience has, however, narrowed the combinations actually used. For example, of the possible power source options, totally DC systems are generally avoided because of arc blow effects although the problem can be reduced by separating the electrodes by more than 50mm. Even when DC is used only on the lead electrode there are occasions when arc blow cannot be avoided as for example when welding inside pipes and in this situation an all AC system must be used. Figures 3.18 and 3.19 illustrate two and three wire systems.

In Europe butt welding is often carried out using a rectifier to supply the lead wire with DCEP. Separate single phase welding transformers are connected to the middle and trail wires, which are spaced 45 and

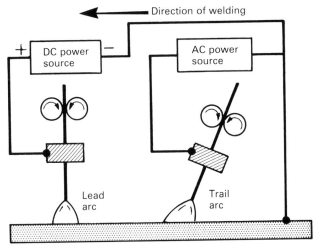

3.18 *Two wire welding.*

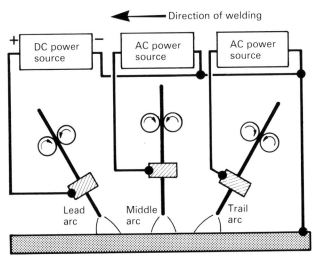

3.19 *Three wire multiple electrode welding.*

65-70mm behind the lead wire respectively. The phasing of the two AC arc voltages is important as magnetic interference between arcs can even occur with fully AC systems unless there is a phase angle difference between the power supplied to the electrodes. The phase difference required tends to be different according to the total power of the system. For this reason separate AC power sources for each wire are often preferred to the Scott type transformer which gives only a 90° phase shift.

At heat inputs of up to 3 kJ/mm (about 15mm plate thickness) the voltage of the leading electrode should be 60° ahead of the trailing one. At higher heat inputs a phase lag of 120° is more suitable. Welding fluxes having a basicity greater than ~ 2 cannot be welded using AC with one wire. With two or three wires, all fed with AC, arc stability is satisfactory when the basicity is less than ~ 3.

Applications

Multipower welding is an important development of the submerged-arc process which is widely used in the linepipe, shipbuilding, pressure vessel and construction industries. Use of several wires gives a marked increase in joint completion rate for fillets and butts, Fig.3.20. For linepipe it is important, because the whole product goes through a single welding station, that high welding speeds are achieved. The linepipe industry therefore makes use of three and four wire techniques.

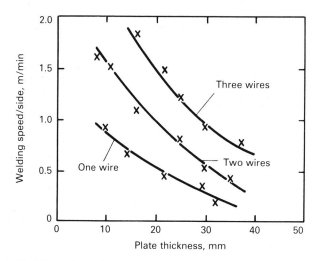

3.20 *Effect of number of electrodes on joint completion rate* (*speed/side*).

Shipyards make extensive use of submerged-arc welding and some employ large specialised plant of which the panel line is an example. The panel line is highly mechanised equipment in which a conveyor system is used to move and support the work along a flowline consisting of a number of workstations. Each workstation is equipped for a specific operation. At the first station the individual plates, accurately flame cut and prepared, are positioned, the long edge joints being faired using a

56

magnetic fairing bed after which they are tacked. The panel may contain up to four plates, each 12-14m in length and up to 3.5m wide, in thicknesses ranging between 12 and 30mm.

At the second station the first sides of the butt joints are welded at high speed using a fixed gantry which spans the panel, and has two separate triple wire submerged-arc heads (usually DCEP/AC/AC) on the same track, see Fig.3.21. Depending on the wire thickness being worked on only two wires may be used. At the next station there is a special high lift crane enabling the panel to be turned over safely in less than 15min, after which it is either returned to the previous butt welding station or progressed to a second butt welding gantry station. In Japan special techniques have been developed for one sided submerged-arc welding so that it becomes unnecessary to turn the plates over.

3.21 *Butt welding station of a shipyard panel line, three wire system.*

Having been welded on both sides the panel is then conveyed to the stiffener injection station. Here the panel stiffeners are delivered automatically, one at a time, to the correct position on the panel and clamped into close contact with it, Fig.3.22. The gantry also supports two twin fillet triple wire submerged-arc heads. Four welding machines work in pairs welding from both ends of the stiffener towards the centre. The system of four machines prevents the fillet welding from becoming a bottleneck in the flowline. The fillet weld throat thicknesses most commonly used for the attachment of stiffeners are 4.5, 5.0 and 5.5mm.

3.22 *Stiffener welding station on panel line — stiffener clamped hydraulically to plate, two three wire fillet welding heads on each side.*

The electrical connection of the three 4mm wires is DCEP, DCEN, AC; arc voltage is between 30 and 40V with welding currents in the range 600-700A. The machines, which work in pairs, are displaced about 100mm relative to each other to avoid magnetic arc blow and to reduce the possibility of porosity from gases trapped between the two fillets.

A system such as that described can produce four panels per day but the integrated handling, fairing and welding installations are extremely expensive, and are economic only when the full production of the shipyard is devoted to large tankers. For plate up to about 25mm thickness a twin wire head may be used as shown in Fig.3.23.

Shipyards are now turning to less specialised, more flexible, techniques such as those used by process plant and construction industries. This equipment is standard with the welding heads mounted on a travelling carriage supported by a gantry which must be substantial to span the panel without significant elastic deflection. In the fillet gantry the welding heads are guided by a jockey wheel inclined to run in the joint ahead of the molten pool, but for butt welding they are usually supported and guided entirely by the gantry as there is often no chamfered edge preparation to guide the jockey wheel. The gantry is usually capable of skewing adjustment across the panel to enable it to accommodate slight errors in alignment. All adjustments and controls are mechanised and operated from a conveniently located pendant control, Fig.3.24.

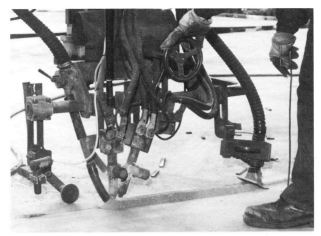

3.23 *Butt welding on the panel line shown in Fig.3.21 using two wires.*

3.24 *Two station plate welding gantry machine with pendant control.*

In the construction industry large non-standard beams are welded in purpose built plant using two twin wire welding heads which make the two fillet welds connecting the web to the flanges simultaneously, Fig.3.25. Tapered beams can be made in special plant through which the tacked beams are fed. Two twin wire heads feeding 2mm wire weld the beams at 1.5 m/min. One head is stationary, the other moves to follow the contour of the beam, Fig.3.26.

Metal powder additions

In conventional SAW, only some 10-20% of the available arc energy is used in melting filler wire, the remainder being dissipated in melting the

59

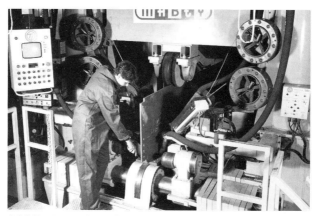

3.25 *Equipment for making two fillet welds simultaneously between web and flange on an I and T beam fabricating machine.*

3.26 *Machine for fabricating tapered beams.*

flux, superheating the molten weld pool and causing fusion of the parent material. This results in the characteristic deep penetration and dilution which is often in excess of that required to ensure adequate weld fusion, particularly for surfacing and the fill runs of a multirun joint. Arc energy expended in superheating the molten pool and causing excessive fusion of the parent material is wasteful but the excess energy is potentially available for melting of additional filler metal in solid or powder form.

Metal powder additions with a submerged-arc were used first several decades ago, and still are, for surfacing where the ability to produce low dilution deposits at high deposition rates contributes to a reduction in costs and an improvement in quality.

60

The first use of a metal (iron) powder addition with a submerged-arc for welding was to increase the thicknesses of steel which could be welded in a single run using conventional techniques. However, the reduction in penetration caused by the powder demanded use of higher currents and/or wider preparations to ensure good root fusion, and this negated the economic advantages of the powder addition. With the development of consumables tolerant to high dilution, joining of thicker section materials could be achieved using joint preparations of low volume, requiring only one or two deep penetration runs to complete the weld. This was a further reason why metal powder additions for welding became unattractive.

More recently, however, an enhancement of the property requirements for the HAZ has begun to limit the usable arc energy so that methods of increasing deposition without raising arc energy, such as metal powder additions, are once again of interest. Use of a metal powder addition in SAW has now become an established technique for welding C, C-Mn and microalloyed steels. It is an inexpensive method of increasing the productivity of standard submerged-arc welding, not only by increasing metal deposition rates, but also through an increase in the melting efficiency of the filler metal, making the technique economically viable in comparison with other high metal deposition rate techniques, e.g. tandem wire SAW. The powder feeding equipment is simple to operate and can easily be attached to both fixed and portable equipment such as a tractor mounted SAW head.

Methods of powder addition

The two most common methods of adding metal powder are illustrated in Fig.3.27 and 3.28. That shown in Fig.3.27 is the forward feed method in which a supply of powder from a metering device, usually a bucket wheel dispenser, is delivered to the joint ahead of the flux. The second method, shown in Fig.3.28, is by magnetic attachment where two or more streams of metal powder, usually metered by passing through a controlling orifice, are directed on to the electrode wire to which they become attached by magnetism to be carried into the molten pool by the wire as it passes through the flux cover.

Metal powder types

For applications in which high toughness is a requirement different types of powder are needed for the two methods of powder addition just described. Early use of the forward feed process often gave

61

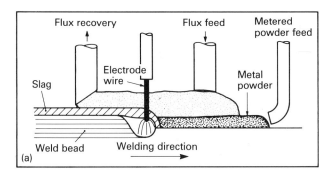

3.27 *Forward feed method of powder addition*: a) *Principle*; b) *In use.*

unsatisfactory weld metal toughness because the powder at that time contained vanadium which was transferred to the weld metal. Powder composition was subsequently modified to match that of the electrode wire, making the weld metal analysis relatively insensitive to the actual powder/electrode wire melt ratio. With the magnetic attachment method the powder is carried directly into the arc region as a result of which transfer of certain elements such as carbon and manganese is reduced. It has been found necessary to add both nickel and molybdenum to the powder to restore weld metal properties.

Metal powders are made by an atomisation process from wires of the appropriate composition but, for special weld compositions, experiments

62

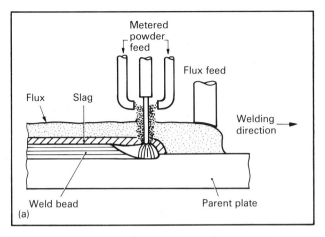

3.28 *Magnetic attachment method of powder addition*: a) *Principle*;
b) *In use*.

have been made with agglomerated powders made from mixtures of iron
and/or alloy powders.

Process characteristics

Powder additions in SAW can be used to achieve any of the three
following objectives:

1 To decrease the number of runs required to complete a joint whilst
   using the same arc energy;

2 To decrease the arc energy used to complete a joint in a given number
   of runs;

63

3 A combination of **1** and **2**.

The first objective is the one most commonly sought and it is the approach adopted by offshore fabricators, who carried out much of the development work in recent years. This is illustrated in the following example. Two joints in 50mm thickness steel plate were welded using a single wire technique and an arc energy of 3 kJ/mm, Fig.3.29. Welding conditions and joint preparation were exactly the same except that in the weld in Fig.3.29*b* a powder addition at a rate of 9 kg/hr was made from the third run onwards using the forward feed technique. The number of runs required was reduced from 31 to 16 when a metal powder was added. There was also a large reduction in the amount of flux used; without a powder addition 7.1kg of slag was produced for each metre of completed weld, whilst with the powder addition, only 3.7 kg/m was generated. It seems therefore that in addition to an increase in joint completion rate, significant savings can be made in the cost of consumables.

The second objective, to use metal powder to decrease the arc energy required to complete a joint in a given number of runs, is rarely required. Indeed the reduction in arc energy which can be achieved is not normally sufficient to give a detectable improvement in HAZ toughness. Generally the requirement is to maximise productivity by using the highest arc energy possible provided the completed weld possesses the necessary mechanical properties. However, the ability to reduce arc energy is relevant to single run weldments.

Many welding procedures which include use of a metal powder addition have been qualified for fabrication of high integrity tubulars for offshore constructions operating in the North Sea, Fig.3.30. Mechanical property requirements are stringent and a minimum Charpy V notch impact toughness of 40J at $-40°C$ is typically specified for the weld metal and HAZ in both the as-welded and post-weld heat treated conditions.

Powder additions can be used to make significant alloy additions to the weld metal such that high arc energy welding produces good weld metal properties. In an actual example, 27J at $-40°C$ was obtained when using an arc energy of 14 kJ/mm for single run welding of a 30mm thickness plate. The HAZ toughness properties were low in the welded joint, however, being only 27J at $+20°C$ in the same weld. This emphasises that to take advantage of this ability to develop tough weld metal at a high arc energy, it is necessary to select a steel capable of giving adequate HAZ toughness.

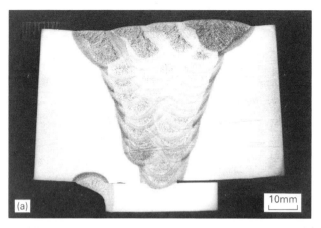

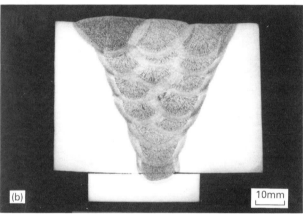

3.29 *Comparison of single sided, single wire welds made with and without powder addition:* a) *No powder, 31 runs in 50mm thickness steel;* b) *9 kg/hr powder from run 3 onwards, 16 runs.*

Forward feed technique

In the forward feed method powder is laid ahead of the weld. Because of the reduction in penetration which results from metal powder additions the technique is not normally used to make single pass welds where the root must be penetrated. If root penetration is attempted the rate of powder addition is restricted. Current industrial practice for multirun single V joints uses the SAW process for fill and capping runs only. The root typically consists of one or two weld runs deposited using a manual or semi-automatic technique. The initial fill runs are then deposited using low arc energy SAW and then the weld is completed using a higher arc

65

3.30 *Forward feed iron powder addition during longitudinal welding of an offshore tubular.*

energy condition. When powder is used it is added in these fill runs. Excess powder in multirun joints can lead to lack of fusion defects but with modest metal powder addition rates, typically 5 kg/hr for single wire welding and 7.5 kg/hr for tandem wire welding, the reduction in penetration does not significantly raise the risk of defects.

In development work on single run welding it has been found that the metal powder buffers the arc and can allow use of a high temperature ceramic backing strip for root run support. The ceramic strip is thermally insulating and restricts heat flow from the root region compared with a copper or steel backing bar. Additional heat is therefore available to cause an increase in penetration or alternatively to melt more metal powder and allow the welding speed to be increased. The reduced penetration with powder was also found to provide a measure of tolerance to variations in root gap. However, for single run welding it is necessary to modify the powder addition rate to compensate for increases in joint volume when the root gap widens.

Magnetic attachment technique

In the magnetic attachment method metal powder is fed on to a slightly extended electrode, typically 55mm stickout, to which it adheres. Some of the powder travels across the arc and some travels around the rear of the arc cavity to be deposited in the tail end of the molten pool, where it is melted by the excess heat in this region. There appears to be no reduction in the amount of parent metal melted, indicating that the excess heat in the molten pool is used to melt the powder. Bead size (parent metal + deposited metal) is therefore increased when metal powder is added using magnetic attachment whereas there is no significant increase in bead size when using the forward feed method of powder addition. An apparently higher melting efficiency with magnetic attachment compared with forward feed allows higher addition rates to be used. With a single wire a typical addition rate would be 9 kg/hr, 80% higher than for single wire welding with the forward feed process.

Increases in productivity for welding flat position fillets can be achieved with both forward feed and magnetic attachment metal powder addition techniques. But the magnetic attachment technique allows the addition of more powder and can therefore deposit larger fillets. It is also applicable to horizontal-vertical (HV) as well as flat fillets, although the maximum leg length of an HV fillet is limited to 10mm.

For grout beading (the deposition of weld beads of specified height and width to provide a 'key' on offshore piling) the magnetic attachment technique is especially advantageous. Powder addition rates are higher than those used for welding, e.g. 14-18 kg/hr instead of 9 kg/hr and grout beads some 9mm high and 25mm wide can be obtained in a single run.

While satisfactory for large diameter circumferential seams the forward feed technique is not practicable on small diameters because the powder must be fed typically 50mm ahead of the arc and would fall away. With the magnetic attachment method, however, the magnetic flux within the parent material is strong enough to hold the metal powder in a tight 'triangular' deposit on small diameter seams but a support may still be necessary for the flux. In this application the magnetic attachment technique is the only practical method of powder addition.

Narrow gap submerged-arc

With all arc welding processes it is necessary when welding material above a thickness of a few millimetres to use a grooved edge preparation to ensure fusion of the root and side wall and to provide access for the arc and electrode. The metal removed for this purpose must then be replaced by weld metal. As plate thickness increases the volume occupied by the edge preparation becomes progressively greater as Fig.3.31 indicates.

The potential economic advantages of narrow gap welding have resulted in much development work over at least two decades in which use of MIG, TIG and submerged-arc processes for this technique has been

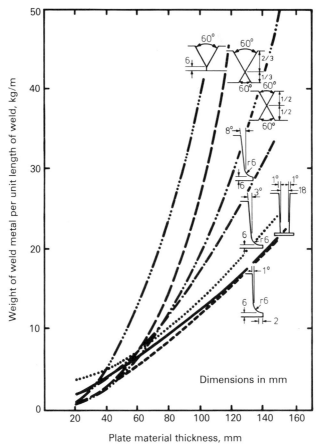

3.31 *Consumption of weld metal for different joint geometries.*

explored. Although submerged-arc possessed a number of inherent advantages, not least the background of experience in conventional welding, it was some time before problems of slag removal and lack of sidewall fusion were overcome. New equipment, consumables and techniques have now been developed and increasing use is being made of the process.

In general, narrow gap submerged-arc (SAW-NG) is applied in heavy section applications over 50mm thickness where single or two sided welding techniques would normally be used. Edge preparation and setting up require precision *e.g.* the gap width variation should be less than 1mm, and applications have tended therefore to be in the pressure

68

vessel field where such controls are acceptable and more readily achieved.

Joint preparation

Edge preparation for SAW-NG depends mainly on which of two possible techniques is to be used. For a single pass/layer the gap is narrow so that both walls of the groove can be fused by the centrally positioned electrode. When there is more than one pass per layer the groove may be wider, Fig.3.32. All grooves have a slight bevel angle of 1-3° to allow for shrinkage and to prevent jamming of the welding head.

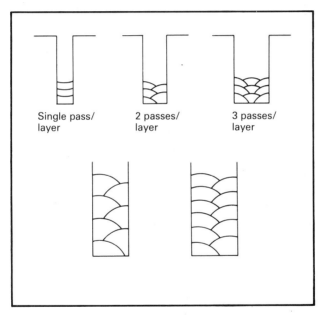

3.32 *Narrow gap welding:* a) *Procedure variations;* b) *Effect of groove width on bead shape.*

The first variant, single pass/layer welding, is capable of welding in gaps down to a minimum of about 14mm, which gives the maximum increase in productivity. The second variant, multipass/layer using two or three passes per layer, does not give the same increase in productivity as a single pass/layer but is significantly more tolerant and more reliable for production of defect-free welds. For multipass/layer welding, the narrow gap weld preparation width is normally about 18mm and over.

69

One pass/layer

For maximum welding efficiency, it is desirable to use the narrowest possible groove and root gap. The single pass/layer technique shown in Fig.3.32a gives the shortest time for joint completion but lacks tolerance mainly because the single wire must be centrally aligned in the narrow gap preparation to achieve constant fusion into both sidewalls. Compensation for irregularities in weld preparation can be achieved only by changing parameters such as travel speed, voltage and current and not by manipulating the electrode wire. The narrower preparations used for single pass/layer welding can also exacerbate slag removal. Experience suggests that for a single pass/layer groove widths should be up to 18mm or slightly greater, with a wire of 3.2mm diameter.

Multipass/layer

Multipass/layer welding often uses two passes per layer as shown in Fig.3.32b. Welding time for two passes/layer is double that for the single pass/layer technique and therefore apparently more expensive. However, it is more reliable and the cost increase is offset by reduced set-up times, fewer delays in production and above all fewer repairs. It allows closer control of heat input and avoidance of slag entrapment which gives sound sidewall fusion and a better bead shape.

The properties of the two pass/layer weld are influenced by the width of the gap which dictates the weld bead shape for a fixed set of welding parameters. Two beads per layer give a better bead profile, more weld metal refinement, narrower heat affected zones and hence superior metallurgical properties. The width of the gap may vary according to plate thickness *e.g.* 14mm for 50mm plate up to 24mm for 300mm thickness plate, but the deciding factor is usually the width demanded by the particular welding head being used.

Welding conditions

Although wire sizes over a range 2-5mm have been used the normal sizes are 3-4mm diameter, the larger size being used for wider gaps in thick plate. Welding conditions are indicated in Fig.3.33 and 3.34. Narrow gap SAW is more tolerant than MIG narrow gap to variations in current, voltage and welding speed. With the single pass/layer method satisfactory welds can be made with arc voltages of 25-30, lower and higher voltages than this giving a tendency to undercutting. Fluxes are selected primarily for ease of detachment of the slag but because a

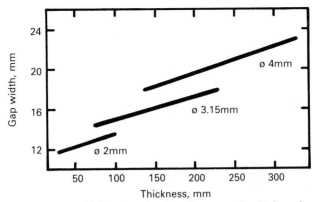

3.33 *Gap width for wires of various diameters, after Malinovska and Pikna.*

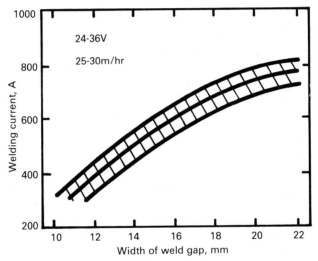

3.34 *Welding conditions for various gap widths, after Malinovska and Pikna.*

multipass procedure is used the flux must not cause a build-up of silicon in the deposits. Although there is some disagreement on the most suitable basicity for the fluxes most operators use neutral or basic fluxes. This ensures clean weld metal with satisfactory notch toughness which is important as applications for the process are in the high quality area.

It is desirable to ensure that the weld beads have a suitable form factor to avoid cracking by selecting the appropriate current, voltage and travel speed. On average penetration depth should equal weld width but deeper welds are possible with low carbon while shallower deposits may be necessary as carbon contents approach 0.2%. Shrinkage leading to a progressive closing of the gap is a problem which can only be minimised by applying heavy restraint to the joint. Circumferential joints are inherently stiff and provide a natural restraint and it may be because of this that many narrow gap joints are of this type.

Welding equipment

Except in Japan, where tandem arc has been used, SAW-NG is invariably carried out with a single electrode wire, possibly because of a desire to keep the process variables for what is a demanding application to a minimum, but it should be possible to use any of the techniques mentioned earlier such as multi-arc or cold wire additions which give an improved deposition rate.

The equipment is basically as illustrated in Fig.3.35 and 3.36. A slightly flattened tube delivers the flux ahead of the wire and its design and the

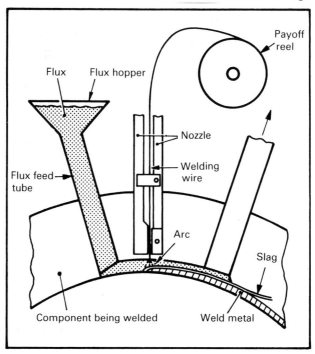

3.35 *Narrow gap SAW.*

3.36 *Narrow gap SAW in use.*

electrode stickout must be such that it does not tend to push the flux ahead in the narrow groove and expose the arc. The construction of the wire guide and contact tube must not allow arc strikes to the sidewalls of the edge preparation. Wire guides are often of hinged plate construction and coated with a thin layer of ceramic for insulation. At the rear of the welding head there is a vacuum extraction unit which removes excess flux. For single pass/layer the wire guide is straight, normal to the weld and disposed centrally in the joint. For multipass/layer a curved or offset guide tube is necessary to direct the wire at the sidewall.

Any automated system for use with SAW-NG must include sensing facilities for maintaining the contact tip at a constant distance from the work despite any workpiece ovality, and also to maintain a constant wire to sidewall distance. For multipass/layer welds, irrespective of the way the workpiece is mounted, means must be included for moving the contact tip from one sidewall to the other on completion of a welding pass and before the start of the next.

Manual override control is often used to allow precise placing of the capping beads. For circumferential welding, anti-drift rotators on the drive for the rolls manipulating the workpieces are indispensable. The system must also be able to maintain a constant welding speed when filling up a deep circumferential groove. This requires that the progression of welding is monitored from the bottom to the top of the joint, the roller bed drive being slowed down as each layer of weld metal fills the groove and the effective welding diameter is increased.

With single pass/layer where the electrode wire is central, arc blow may not be a particular problem and DCEP has been used. With a multipass/layer technique it has been reported that arc blow is not encountered when making circumferential welds. Alternating current, however, gives complete freedom from arc blow and particular advantages are claimed for square wave AC. This is even less influenced by magnetic effects and it gives better arc stability, allowing more freedom in the constitution of the flux.

Applications

Submerged-arc narrow gap welding is apparently gaining in popularity although at the present time there are more examples of use of the competing method — gas metal arc (MIG) narrow gap. A particular claim for SAW-NG is that it seems to be more tolerant to welding conditions than the MIG version. The minimum thickness at which the process becomes economically attractive depends to a degree on the quality required in terms of lack of defects and weld properties. Because SAW-NG is a closely controlled continuous process which eliminates stops and starts and associated defects it offers high quality welds. For work where HAZ toughness is important the minimum thickness for economic welding may be as low as 25mm because the deposition rate of conventional SAW techniques would be restricted by heat input limitations.

In deciding whether or not to use SAW-NG many factors must be considered. The equipment is expensive and operators require special training. Any gain in deposition rate must be compared with gains which might be made by using tandem welding or addition of metal powder. Edge preparations must be machined while for other SAW methods gas cut edges would probably suffice. Means must be found for forming the root weld or removing any backing bar which might be employed. Finally, a realistic strategy must be worked out to deal with any defects which may be found.

The main application area is in heavy fabrication particularly circumferential welds for the power plant industry. Both alloy steels and stainless steels have been welded successfully.

CHAPTER 4
Consumables

Unlike the electrode for manual metal arc welding in which the wire and flux combination and the relative melting of each are fixed and beyond the control of the user, consumables for submerged-arc welding can be combined to the wishes of the user and fused in varying ways to give different weld metal compositions. The features of the process which call for special consideration as they affect the selection of consumables, *i.e.* the choice of combination of flux and electrode wire, are as follows.

High dilution

High dilution of the parent plate into the root passes, which are the only passes in single and two pass welds, considerably reduces the ability to alter weld composition by changing the wire. With 70% dilution (not an unusually high value) changing from a 0.5 to a 2%Mn wire increases Mn input to the weld by only 0.45%, but most fluxes influence Mn transfer in such a way as to reduce this to a difference in weld Mn content of perhaps 0.3%.

Energy input

The high productivity of submerged-arc welding arises from the possibility of using high welding currents and therefore high energy input. This leads to low weld cooling rates and hence a tendency for low strength in the weld and low toughness in both weld metal and HAZ. For certain high quality applications such as offshore structures a limit (5 kJ/mm) may be placed on the process which can have serious cost implications.

Dependence of weld composition on flux type

Some fluxes are Mn-alloying; this category includes not only the manganese silicate fluxes but some other types, particularly alumina fluxes. High silica fluxes (including manganese silicate types) increase the weld Si content, decrease its carbon content, and give a high weld oxygen content and hence a high inclusion content. Basic fluxes may add little or no silicon and remove little carbon but they give low oxygen contents which, combined with their ability to reduce all except the lowest sulphur levels, lead to low inclusion contents. Most fluxes, unless they are specially formulated, tend to add slightly to the phosphorus content of the weld and chromium is lost unless a compensating flux is used.

Economy of welding

The major features needed for good economy are a high speed welding capability or a high deposition rate, good slag detachability and in some circumstances the ability to weld on to rusty plate. These features are principally controlled by the welding flux, as can be seen from the advantages and limitations set out in Table 4.1. Economy is seriously affected by delays on the shop floor, a common cause of which is defects awaiting repair. An unsuitable flux can increase the risk of defects and be a contributory factor.

Bead shape

A satisfactory bead shape is essential in many applications, such as pipe welding, to avoid harmful stress concentrations which would otherwise require expensive machining operations for their removal. Bead shape is controlled by both welding parameters and certain properties of the molten flux or slag; probably viscosity and surface tension, although the relationships are not at all clear.

Weld and HAZ properties

Of the commonly required mechanical properties weld metal toughness is the most complex to achieve, partly because it is controlled by many factors and partly because it comprises resistance to two types of fracture; brittle cleavage and ductile microvoid coalescence. Brittle fracture resistance is controlled by the weld microstructure and its yield strength, which are largely governed by the weld composition and its cooling rate. Here the flux is only one of the factors which must be considered along with plate and wire compositions and welding parameters. Although soft weld metals with a coarse pro-eutectoid ferrite microstructure, Fig.4.1a, usually have adequate toughness, this phase as well as coarsely bainitic (or ferritic side plate) structures, Fig.4.1b, should be avoided for higher yield strength weld metal. Here a fine acicular ferrite microstructure as in Fig.4.1c is desirable. This can be achieved with elements which increase the hardenability of the weld metal, such as Mn, Mo, Ni and B. Micro-alloying elements in the parent steel, such as Nb and V, and Al can help to

Table 4.1 Methods of describing submerged-arc fluxes

a) By composition

Main types	Advantages	Limitations
A Calcium silicate, high silica	High welding current, tolerant to rust	Poor weld toughness
B Calcium silicate	Moderate strength and toughness, multipass welds	
C Calcium silicate, low silica	Good toughness with medium strength, fast speeds	Not tolerant to rust, not used for multiwire welding
D Basic (low silica, mid-alumina)	Good strength and toughness in multipass welds	Not tolerant to rust, limited to DC positive welding, poor slag detachability
E Manganese silicate	Moderate strength and toughness, tolerant to rust, fast welding speeds	Limited use for multipass welding
F Alumina (bauxite based)	Fairly high currents, fast welding speeds, good toughness of two pass welds, tolerant to rust	Limited use for multipass welds or when stress relieved
G Alumina rutile		

b) By chemical characteristic

Main types	Compositional types (see above)	Main characteristics
Acid	A, E	High weld oxygen content, increase in Si on welding, loss in carbon
Neutral	B, F	Less change in weld composition and lower oxygen than for acid type
Basic	C	Less change in composition and lower oxygen for basic type
Highly basic	D	No change in carbon, loss of sulphur and Si (except from low levels)
Manganese alloying	E (C)	Adds Mn to weld, other changes depend on basicity

c) By method of manufacture

Fused	All except D	Low moisture content and pick-up, cannot include additions which decompose or oxidise during fusion
Agglomerated*	All	Less severe limits on formulation set by drying temperature

*Frequently referred to as 'bonded' in the USA and 'ceramic' in translations from Russian

Table 4.1 contd Methods of describing submerged-arc fluxes

d) By use

Main types	Compositional type (see above)	Main characteristics
Multipurpose	B	One, two, or multipass, all current types, moderate strength and toughness, tolerant to rust
General purpose	A,E	Used where no toughness requirement, tolerant to rust
Special purpose, high speed	F, C	High speed welding possible with moderate to good toughness
Multipass	All except E and some C	No build-up of Mn during welding
Notch tough	C, D, F	Good toughness, not all types suitable for two or multipass welding or stress relief

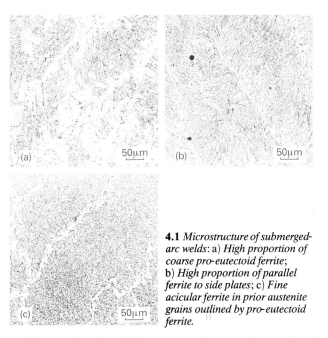

4.1 *Microstructure of submerged-arc welds*: a) *High proportion of coarse pro-eutectoid ferrite*; b) *High proportion of parallel ferrite to side plates*; c) *Fine acicular ferrite in prior austenite grains outlined by pro-eutectoid ferrite.*

improve weld metal hardenability and toughness when present in small quantities. In excess they can reduce toughness by promoting bainite formation, by encouraging age hardening (especially during stress relief heat treatment), and also by giving fine martensitic phases, Fig.4.2, which

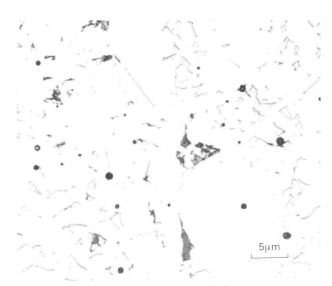

4.2 *Fine dark and light martensitic phases revealed by etching in picral.*

are particularly harmful in heavily banded and segregated steels. High Nb with Mo is particularly undesirable if the weld is to be heat treated for stress relief. Lower oxygen contents than are given by the current basic fluxes may not be helpful as they can give rise to welds containing an excess of aluminium (normally diluted from the plate) over oxygen. Such an excess alters the microstructure and seriously impairs toughness.

Above the transition temperature, good toughness involves a good resistance to microvoid coalescence which requires a low content of inclusions in the weld, *i.e.* low O and S levels. These are best achieved with a basic flux, which gives low weld oxygen content in the range 0.02-0.03% together with a desulphurising ability, but it is also helpful to start with low S plate and wire. For many applications, however, fluxes of the alumina and low silica calcium silicate types are adequate.

Other things being equal the transition between brittle and ductile fracture depends on the resistance of the weld metal to these two types of fracture and also on the way in which its yield strength varies with temperature. In practical terms a requirement for a low transition temperature may be fulfilled by the presence of adequate alloying elements in the weld but if a high Charpy or CTOD value is also required, a clean weld metal with low O and S contents is also necessary.

The HAZ toughness depends on the composition of the parent steel, the energy input of the welding procedure and whether or not a stress relief heat treatment is used. Weld strength is normally easy to achieve with the available wires and fluxes. Some difficulties may be encountered in high strength low alloy steels above say 500 N/mm^2 where some restriction of energy input may be needed. Welds of somewhat lower strength, 350 N/mm^2 and above, which require normalisation also need careful selection of consumables if matching strength or better is required.

Avoidance of cracks and other defects

This subject was discussed in Chapter 2. As far as hydrogen cracking is concerned use of clean wire and dry flux (particularly when the agglomerated type is used) to minimise potential hydrogen input to the weld is most important. Solidification cracking depends on the composition of the weld, its solidification pattern and the strain on the solidifying weld metal but only the composition is directly affected by welding consumables. As was discussed in Chapter 2 carbon is potentially the most damaging element which, if other considerations allow, can be kept low in the weld by use of high silica fluxes, *i.e.* Mn and calcium silicate types.

Trapped slag is usually found only in multirun welds and can be avoided by selecting a flux designed for this type of welding and ensuring that welding conditions give a smoothly blending toe to the weld and that the beads are placed to avoid creation of deep narrow gaps between the beads and fusion faces in which slag might become trapped.

Types of flux and their development

There are numerous classification schemes for flux, some of which are outlined in Table 4.1. Perhaps the most straightfoward approach is the simple compositional scheme outlined in the first part of the Table. Provided that distinction is made between calcium silicate types the categories are fairly well defined and other types can be added as necessary. Chemical characteristics and manufacturing methods are too simple to provide a definitive classification (Table 4.1, parts *b* and *c*) although it is useful to know if a flux is fused or agglomerated and whether or not it is manganese alloying as are some alumina fluxes. Description by use is often of limited help to the more advanced user because several compositional types may fall into one usage type. Other systems based on letter and number codes appear to be too complex and difficult to learn. There is a considerable difference between the classification schemes of the various international standards.

For the average user the flux is selected on the basis of factors such as the need to weld at high speed, to weld over rust or scale, to make multirun welds or perhaps for the flux to be stable on small diameter workpieces. Only with heavier fabrications or those made to exacting specifications is it necessary to give priority to the characteristics of the flux and wire combination that affect weld properties, particularly toughness.

The high silica, calcium silicate type now obsolescent was probably the first successful variety of submerged-arc flux to be developed and could be used to make large weld beads with high currents. Unfortunately it gave welds of poor toughness and could not be used for multipass welding because of a high Si pick-up and cracking problems.

A range of fluxes was then developed containing less silica and more lime (CaO) which is preferably added (at least to the agglomerated fluxes) as calcium silicate because lime absorbs water from the air too easily. To reduce the silica content below that of calcium silicate without adding lime other compounds are used such as magnesium oxide, calcium fluoride and alumina (already present in smaller amounts in the calcium silicate fluxes).

Outside this mainstream of development lie manganese silicate fluxes, suitable for high welding speeds and alumina fluxes which have the additional advantages of high current-carrying capacity while giving lower weld oxygen content than the manganese silicate types. In addition some fluxes intermediate between the main types are available as well as a rutile type containing boron.

Development of consumables for submerged-arc welding was greatly encouraged by the offshore construction industry which placed stringent requirements on weld toughness. Initially these requirements could be met only by use of highly basic agglomerated fluxes, DCEP and single wires with an uneconomic low heat input. Later, deposition rates were increased by using two wires in tandem. More recently developed fluxes offer better properties but only by careful selection of the flux/wire combination. For the highest toughness properties where high dilution occurs basic fluxes with wires depositing welds containing titanium and boron are required.

Wires

The main classification of wires is by Mn content and wires containing 0.5, 1.0, 1.5 and 2.0%Mn (S1-S4) are available to give welds of progressively increasing strength. It is necessary to select the wire in conjunction with the flux as the latter affects Mn transfer during welding.

The strongly Mn alloying manganese silicate fluxes normally require only a low Mn, S1 wire, whereas the high silica calcium silicate type may require an S4, 2%Mn wire.

For extra hardenability 0.5%Mo versions of most of the Mn levels are available. Besides giving extra strength, the extra hardenability can promote improved toughness in many instances as discussed later. Both Mo-bearing and Mo-free wires may contain 1%Ni added to improve cleavage resistance and low temperature toughness. There are also Mo-Ti-B wires formulated to deal with high dilution and chromium and Cr-Mo wires for welding Cr and Cr-Mo steels.

Most wire grades, but especially the S3 types, are available with low S and P content. This helps to confer a low inclusion content which is useful in achieving high Charpy values and maintaining good resistance to shear fracture as well as reducing the risk of solidification cracking. Such wires are sometimes described as specially deoxidised. Highly alloyed flux-cored wires have been developed for 2-4 arc multi-arc welding for use with the cheaper simple S1 wire to provide the required weld metal composition.

Flux/wire combination

As it is the combination of flux and wire which determines the composition of the weld metal, other things being equal, it might be expected that alloying additions could be made equally well either way. From the practical point of view, however, it is usually preferred to use simple standard wires and make whatever additions are necessary through the flux. This is mainly because the economical production of wires necessitates reasonably large heats of steel and special compositions are therefore difficult to obtain. In contrast, special flux formulations are readily made in smaller quantities. There is also resistance in many large workshops where a variety of work is carried out to having too many wires and fluxes because of possible mistakes on the shopfloor. There are usually several ways of achieving desired welding characteristics and weld properties. However, once a selection has been made and the welding procedure has been worked out changes should only be made after careful consideration. Most national standards work on the basis of flux/wire combination to give specified weld properties and they and the consumable supplier's information should be consulted when deciding which flux/wire combination to use. National standards are referred to in Chapter 1.

Many of the recently developed consumables for exacting applications contain boron which is capable of giving good weld toughness in terms of cleavage resistance as a result of the increased hardenability it confers. Boron may be added via the flux, alloying in a solid wire, or via a cored wire. There are problems in achieving efficient and consistent transfer of the element which needs to be protected by a strong deoxidant (usually titanium). Alloying is more efficient if a basic flux giving a low weld oxygen content is also used.

A range of C-Mn-Mo wires with Ti and B additions is now available and is used when the highest weld metal toughness is required. The Ti-B consumables provide improvements in both upper shelf energy and transition temperature. Cored wires offer an efficient way of introducing a range of elements into the weld metal and fully basic cored wires are in common use in some countries although less so in the UK.

The ability to deposit tough weld metal confers greater freedom in the heat inputs which may be used, however, HAZ properties may then have to be considered. Various techniques for increasing the deposition rate without raising heat input, *e.g.* multiwire welding or metal powder additions, are discussed in Chapter 3. Whatever mechanism is used to achieve the desired weld metal analysis and structure it is the total consumable concept, flux, wire, cored wire or metal powder which must be considered.

Consumables for different steel types

Mild steel

Except for low temperature applications the choice of flux depends essentially on the weldability required. For low temperature applications, particularly where resistance to dynamic fracture is necessary, a flux giving a clean weld metal of low inclusion content is indicated. Stress relief has little effect on toughness and reduces strength only slightly. Some care is needed in selection of consumables for welds which require normalising as the low C content of mild steel weld metals made with most types of flux results in low strength and toughness after normalising. A basic, alumina, or low silica calcium silicate flux coupled with an alloyed wire (suitable combinations of Mn, Mo and B have been proposed) is probably best.

C-Mn microalloyed steels

Pick-up of microalloying elements in high dilution runs has a considerable effect on weld properties. Niobium strengthens the weld but

84

may reduce toughness, particularly if the weld hardenability in terms of Mn and/or Mo content is inadequate. Particular care is needed if stress relief is used because Nb and Mo can together seriously reduce toughness. Vanadium also strengthens the weld but reduces toughness if in excess, particularly if Nb is also present. Rare earth metals probably have little effect.

To achieve the required toughness with strength, adequate weld hardenability is needed and this depends on sufficiently high Mn with Mo, Ni and/or B selected according to the microalloying elements, and the need for heat treatment. Excessive alloying raises the yield strength too much and again reduces toughness. A limited weld bead size may also be required to ensure adequate hardenability. For good resistance to ductile fracture, *e.g.* if a high Charpy value is demanded, then alumina, low silica calcium silicate, or basic fluxes are needed to produce weld metal of the necessary low inclusion content.

Alloy steels

An alloy steel wire is required, coupled with a restricted energy input and a flux selected from the three groups mentioned at the end of the previous section. For the highest strength levels the best combination of strength and toughness requires a low carbon martensitic microstructure. At such strengths a good resistance to ductile fracture assumes increasing importance and hence the particular need for a clean weld metal. With stainless steels, wires of a matching composition are usually employed and the flux is compounded to ensure that there is no loss of alloying elements.

Flux delivery systems

To reduce moisture contamination and wastage, and create a better working environment, mechanised flux feed and recovery facilities are now used extensively. Although these systems are favoured for high integrity applications or when applying submerged-arc in a confined space they make good economic sense in the normal workshop. Flux delivery and recovery systems are of two main types, small portable units used alongside the welding machine and larger fixed installations, remote from the welding machines.

The small portable units are mounted on the flux delivery system on the welding machine. Dust and fines are cleaned from the recovered flux and it is fed to a container above the normal flux hopper into which it is discharged manually from time to time. A 3m vacuum hose connects the unit to the pick-up nozzle, see Fig.2.3.

With large fixed installations use of pressure vessel storage with a minimum level sensor facility allows flux to be fed with compressed air to the welding head. The return is made through a vacuum unit and a centrifugal filter before storage in an absorber vessel before re-entry into the main pressure vessel. Capacities of such systems vary depending on feed distances, which may be up to 20m, and process requirements. It is common to feed two or more welding heads and to have extra facilities such as fresh flux addition and/or a heated pressure vessel to ensure drying and optimum storage of the flux, Fig.4.3. When compressed air from a workshop line is used to transport flux it is important that the air is dry and that there are suitable moisture traps in the system.

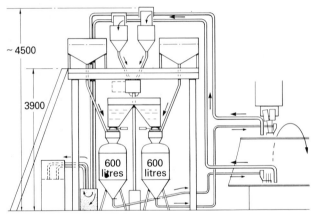

4.3 *Typical force flux feed and recovery system as used for internal and external welding of linepipe.*

CHAPTER 5
Welding procedures

A welding engineer contemplating use of submerged-arc must consider two issues, weld quality and production cost, which are closely interrelated. This chapter examines factors which the welding engineer must take into account and aims to offer guidance on how to make a judgement on using the process. It does not provide a catalogue of welding conditions, not just because of the sheer volume of such information but also because equipment and consumable manufacturers and published information alike quote a range of possible welding conditions. Figure 5.1 summarises the relationship between usable current range and wire diameter as found in published sources. The

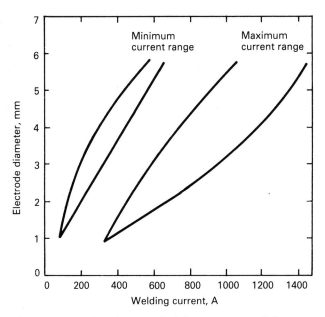

5.1 *Variations of maximum and minimum recommended currents for a range of submerged-arc electrode diameters.*

87

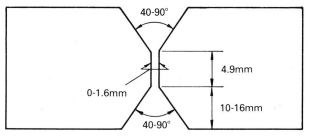

5.2 *Variations in recommended joint designs for two pass welding of 25 mm plates.*

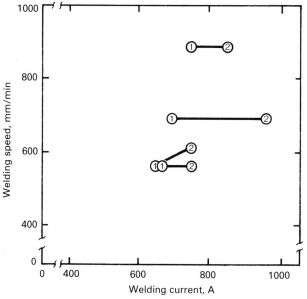

5.3 *Range, 1-2, of reported values of welding conditions.*

amount of disagreement is clear, and this range of opinion is also seen in the recommendation for joint preparation summarised in Fig.5.2 and welding parameters for a simple application plotted in Fig.5.3.

Although these figures illustrate the inherent tolerance of the submerged-arc process, they also have important implications with regard to production rates and consumable usage. If a range of welding techniques or procedures is equally feasible for a given welding task, it follows that some must be more productive than others and it is important that selection should be made on the basis of economics.

88

Welding costs

The major factors

There are two principal factors which affect the cost of welding — duty cycle and deposition rate — and they are applicable to all arc welding processes. There are in addition costs incurred in maintaining the required quality. Duty cycle is defined simply as the ratio of arc time to employed time and deposition rate is the rate at which weld metal is deposited.

Duty cycle depends not only on the characteristics of the welding process but also on a number of other factors such as support operations, *e.g.* cranes and material supply, supervision, the organisation of work and operators, and the reliability, design and ease of maintenance of equipment. The effect of duty cycle on the cost of deposited weld metal is significant, hence the need to achieve a high duty cycle not only for reduced floor time but for lower costs. Within reason, the simpler the equipment the more acceptable it is to the welder and the easier it is to maintain, all of which contributes to a higher duty cycle.

Welding efficiency can be improved either by decreasing the amount of weld metal required or increasing the rate of deposition or a combination of both. To decrease the amount of weld metal required the narrow gap welding technique has been developed which is discussed in Chapter 3. It is a specialised technique, however, which has yet to reach maximum potential in the heavy steel fabricating industry. The most widely used approach to reduce costs has been to increase the weld metal deposition rate. Deposition rate depends basically on the welding current passing through the electrode(s) and may be increased by raising the current or by adding extra metal for the arc to fuse, *e.g.* hot and cold wire and metal powder additions.

The submerged-arc process has the ability to capitalise on duty cycle and deposition rate. It has the potential for high duty cycles as it is usually in a mechanised form and it also has the characteristic capability to carry high currents and hence can give high deposition rates. Sometimes, however, efficiency gains are more subtle as, for example, when the process is used in applications, particularly with stainless steel, where surface finish is important. The smooth finish and absence of spatter obtained with submerged-arc may make expensive surface grinding and finishing operations unnecessary.

89

Raising productivity

All attempts at raising productivity should begin with a thorough appraisal of the whole manufacturing process, both before and after the welding stage. Factors such as material and work flow and the amount of rework because of defects should be examined and particular note taken of any bottlenecks in production. Only when this has been done should attention be turned to the details of the welding process.

In attempting to increase the effectiveness of the submerged-arc process itself fabricators have turned to high welding currents and multiple arc systems. However, although use of high currents or multiple arcs may increase deposition rate it sometimes does so at the expense of weld metal toughness, or at the risk of centreline cracking, hot cracking or poor slag detachability.

Increasing deposition rate without an appreciable increase in welding current by using such techniques as hot or cold wire addition, long stickout, or electrode negative polarity also has its limitations. Although there are many successful applications, these methods can give rise to arc starting problems and lack of sidewall fusion and may not be readily accepted by shopfloor personnel.

Other more passive methods of speeding production have potential. Use of ceramic backing strips can reduce cooling rate and allow speed to be increased for a weld of the same size. Back gouging, an expensive procedure, can be eliminated sometimes by using the 'punch through' method. After superficial cleaning of the second side of a butt joint, a deeply penetrating bead fuses into the root run on the first side. There is high dilution, however, and where weld metal toughness is a crucial factor special consumables, *e.g.* Ti-B containing wires and fluxes, may be necessary.

Equipment development

Changing attitudes on the part of both equipment manufacturers and users have now resulted in more widespread use of plant which has increased duty cycles and reduced labour involvement. Use of automated and work handling systems, which allow maximum use of welding in the flat position, has been extended. Gantry systems have been fitted with seam tracking using tactile or inductive sensors to maintain vertical and lateral alignment. Remote video observation and control is now available from well designed control pendants. These may include motorised cross slides operated by a joystick type control for positioning the welding head, Fig.2.4. Together with microprocessor control and pre-programming these facilities make it easier to set and maintain precise control of welding parameters. There is also now a greater availability and use of simple mechanised systems using lightweight compact tractors with simple devices to steer these self-propelled units along the joint. Improvements in flux delivery and recover (see Chapter 4) have also contributed to shopfloor efficiency.

Establishing a procedure

General

When defining welding procedures for particular applications, the welding engineer must first consider the specific weld qualities required, *e.g.* penetration depth, weld surface profiles, mechanical properties, *etc.* The relevant welding parameters expected to produce a weld meeting these standards are then selected. In theory there is direct control over a large number of welding variables. In practice, however, although some factors can be fixed quite accurately others vary during the course of fabrication. Outside limits can be established for such variations but precise settings cannot be maintained. The dilemma to be faced in this situation is selection of welding conditions which combine tolerance to uncontrolled variations in parameters with the highest output rate while still producing an acceptable weld.

Welding conditions selected on the basis of experience may well give satisfactory results but there is no guarantee that optimum production rate and process tolerance are being achieved. Such assurance can result only from a background of quantitative process data. The Welding Institute formulated 'procedure optimisation' as a method of supplying such process information and this chapter introduces this approach.

The variables

The first step in procedure optimisation is to list the process variables. These are particularly numerous in submerged-arc welding and may be divided into classes depending on the ease with which they can be varied in practice. The major factors affecting the welding process are:

1 Preset variables, *e.g.* material grade, material thickness, design, weld property requirements. These variables are usually outside the control of the welding engineer, forming part of the fixed requirements of the customer.

2 Background variables, *e.g.* basic technique, (single arc/multipower, *etc*) equipment type, flux/wire combination, wire diameter, edge preparation design. These factors are decided at an early stage of production, either as a result of outside purchasing decisions or the availability of welding plant. They will, however, determine general production capability and economics.

3 Secondary variables, *e.g.* contact tip/workpiece distance, electrode angle. These parameters are set at normal values as part of the detailed welding procedure, but minor variations encountered in practice are unlikely to affect weld quality.

4 Primary variables, *e.g.* current, arc voltage, welding speed. Once the basic procedure defined by the preset and predetermined variables is decided these welding parameters are of particular significance. The levels set in the production procedure and the variations encountered in practice directly affect weld quality and output rate. When the welding speed is high factors such as irregularities in plate preparation and cleanness as well as fluctuations in power and speed have a proportionally greater effect.

The procedure optimisation approach is particularly concerned with the magnitude and importance of variations in welding parameters which can arise in production. These can originate in several areas and include those which follow.

Instrumentation

The instrumentation necessary depends on the type of submerged-arc technique being used and the importance of the application. When using a self-adjusting arc with a constant wire feed speed/constant voltage power source it is important to check both wire feed speed (which determines welding current) and arc power source voltages. Electrode extension (which affects burn-off rate) should also be monitored. With voltage control systems using constant current power sources wire feed speed is continually varying and is not normally measured but its effect is seen in the welding current, measurement of which becomes important. Modern power sources allow correction of line-voltage variations which greatly assists maintenance of correct welding variables from day to day.

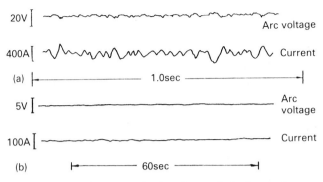

5.4 *Recordings of current and voltage for a typical submerged-arc weld*: a) *High speed recording*; b) *Recorded using highly dumped instrument.*

Welding current and arc voltage are the most important parameters which can be continuously monitored during submerged-arc welding. High speed current and voltage recordings, Fig.5.4, reveal considerable high frequency fluctuations whose origin lies in the processes of metal transfer, slag cavity pulsation and molten pool movement, coupled with interaction with power source and wire feed characteristics. As these fluctuations are short term, a given welding condition characterised by mean values of current and arc voltage is capable of producing reproducible weld beads. This type of monitoring is provided by highly damped metering systems which smooth short term fluctuations and make recognition of significant procedural changes easier, Fig.5.4b.

The success of such a system depends, however, on the accuracy of the instruments or even, as sometimes happens, whether or not they work at all. In general, well-damped instruments on welding units cannot be read to an accuracy of better than ± 2%. In addition to potential reading errors, there is always the possibility of instrument error through misuse, infrequent calibration, or, with voltmeters, voltage drops in the welding circuit. For this reason it is doubtful if present units could be expected to agree with each other to an accuracy better than ± 5%.

Machine characteristics

As mentioned above the characteristics of the power source and wire feed control unit affect primary control parameters at a basic level. These high frequency effects will be smoothed to some extent by inertia, but may be seen in variations in bead quality and process capability at the extremes of the process range.

Preparation and fit-up

A given production route presents the welding engineer with joint preparations whose critical dimensions (angle, root face, gap, mismatch) vary within limits. Although the exact limits of variations are often unknown the outside extremes can usually be specified and used in procedure calculations. Smoothness and cleanness of edge preparations are more critical the higher the welding speed.

Consumables

Dimensional tolerances on welding wire are tight and do not usually present problems in submerged-arc welding. Wire composition, surface finish, cleanness and straightness are occasionally sources of difficulty, but these factors are covered by good housekeeping. Welding flux can sometimes cause procedural problems, especially if drying operations are omitted or there is uncontrolled recycling of flux resulting in damp, contaminated or mixed flux. This also is a matter of good housekeeping.

Operator variables

In spite of the high level of mechanisation attained in many submerged-arc applications, the welding operator remains one of the most vital links in the production chain. Without conscientious observation of welding procedures by the operator there is no hope of maintaining production efficiency.

When setting up a welding procedure decisions are faced regarding process variables more or less under the control of the welding engineer but all of which affect weld quality and productivity. The success achieved in setting individual procedures depends on:

— The welding engineer's and the company's experience;

— Published information;

— The success with which the above are transferred to production.

Procedural options

Procedural options open to the welding engineer can be visualised by considering a simple butt weld, Fig.5.5. For weld geometry, acceptable limits can be defined for penetration, bead width, and bead height, and the range of limits may be extended to include penetration/bead width and bead height/bead width ratios which characterise weld geometry. As an example, the limits which may be applied to two pass square edge butt welds in 12.7mm steel plate are shown in Table 5.1.

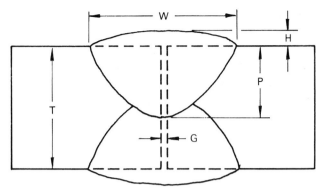

5.5 *Two pass butt weld in 13mm steel, basic geometric parameters and ranges of conditions, see Table 5.1.*

Table 5.1 Basic weld quality criteria for two pass square edge joints in 12.7mm steel plate (for notation see Fig.5.5)

Penetration, P	Bead width, W, mm	Bead height, H, mm	P/W	H/W	Undercut
55% ≤ P ≤ 80%	W ≤ 20.0	H ≤ 4.0	P/W ≤ 0.75	H/W ≤ 0.30	Absent

Note: Sudden changes in thickness are generally undesirable and it is sometimes required that any change in thickness should not occur over a slope of more than 1:4. In the example quoted, at the extreme limit the increase in thickness is twice 4mm which would require a bead width of not less than 32mm. As this weld width may be considered excessive it might be necessary to restrict the bead height to not more than, say, 3mm which would require a weld having a width of a minimum of 24mm.

For a given type of welding unit, and with fixed consumables and plating tolerance, variation of bead quality depends on the values of the main control variables (arc voltage, current and welding speed). At a given voltage, combinations of current and welding speed either satisfy, or fail to meet, the weld quality criteria and a process field may be constructed as in Fig.5.6. Viable procedures are restricted to the interior of the process field. Welds outside this region are defective and particular types of defect may be identified with individual border lines, as in Fig.5.7. This type of construction may be extended to include the effect of voltage, Fig.5.7*a* and superimposition of fields at different voltages allows assessment of voltage tolerances, Fig.5.7*b*.

Although these examples are based on weld geometry criteria there is no reason why the range of quality limits cannot be extended to include other

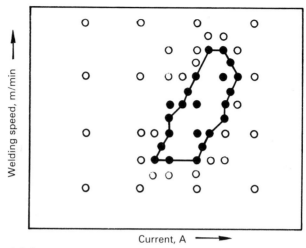

5.6 *Locating a process field.*

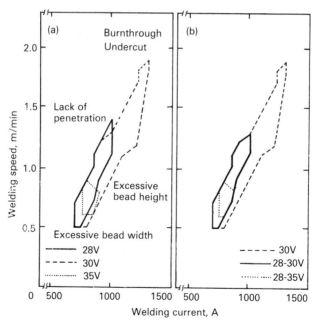

5.7 *Procedural fields for a square edge butt in 12.7 mm steel, DCEP, fused acid flux, 5 mm diameter wire:* a) 28, 30 *and* 35 V; b) 28-30 V *and* 28-35 V.

96

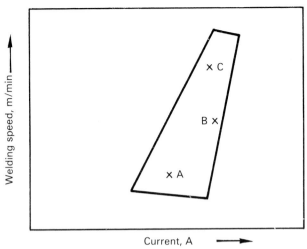

5.8 *Production situations from the process field viewpoint.*

factors, including metallurgical and mechanical property requirements. Some standards of weld geometry will, however, always form the basis of any procedure optimisation approach. The concept of a process field containing all viable procedures for a given application accounts for variations found in welding practice, and can be used to explain reasons for varying degrees of success in production. Figure 5.8 in simplified form shows possible situations:

(A) The procedure is tolerant and production is uneventful. Productivity is, however, low. This situation is perhaps the most common in normal production. Without quantitative data to prove the possibility of improved productivity, changes to established procedures can be extremely difficult to implement;

(B) The procedure is close to a boundary where defects may occur. Small variations in one or more parameters may give intermittent defects which can become extremely troublesome;

(C) The procedure is well balanced, demanding control tolerance within the capacity of the equipment and offering good productivity. The objective of procedure optimisation is to select welding conditions within the last category.

A range of welding conditions could be chosen for given applications and three main types of condition may be identified, Fig.5.9. These are tolerant, economic, and optimal procedures which offer progressively increasing levels of productivity as process control is increased.

97

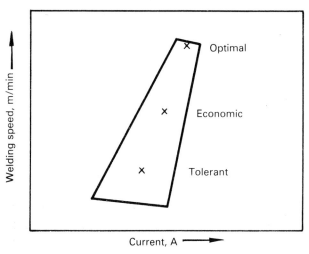

5.9 *Nominal welding conditions.*

Application and uses of optimisation

Optimisation involves the following steps:

1 Definition of production objectives, *e.g.* high speed, good surface finish, specific compositional requirements, heat input limits, *etc*;

2 Recognition of potential defects and other production problems, *e.g.* cracking, porosity, distortion, which may place limits on a procedure;

3 Recognition of control limits in production, *e.g.* equipment variability, instrumentation, joint preparation and fit-up tolerance.

Once objectives of optimisation and control tolerances available are known the link between optimised data and the production situation can be made. The success of such an operation depends on the levels of control available: at the very least welding conditions generated will form a good basis for shortened production trials. At the other end of the scale improved control equipment could allow more direct transfer of conditions from potential to practical use.

Process data of the type produced in optimisation studies may be used in four ways:

1 Selection of production procedures to allow more efficient use of existing equipment;

2 Comparison of alternative procedures using different joint preparations or process variations, *e.g.* single electrode versus

98

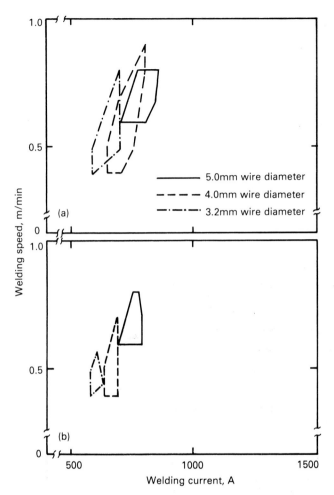

5.10 *Effect of wire diameter and bead shape criteria on procedural fields for 12.7 mm steel plate welded at 28-35 V, DCEP with fused acid flux*: a) *Penetration/width* < 0.75; b) *Penetration/width* < 0.60.

multiwire. Figure 5.10 shows process fields for square edge butt welding of 12.7mm steel plate with 3.2, 4.0 and 5.0mm diameter electrodes. With a penetration/bead width limit of 0.75, the fields are of approximately equal area. Specification of a penetration/bead width maximum of 0.60, Fig.5.10*b*, reveals a marked reduction in process tolerance for the smaller diameter electrodes. This feature could be of importance if plate carbon levels are high and the danger of

99

solidification cracking makes such a weld bead geometry restriction advisable;

3 Equipment and consumable development to improve process tolerance and extend process capabilities. Figure 5.7*b* shows that increases in joining rate of up to 40% could be obtained for equivalent weld quality and with a tolerance of 25A, by controlling the welding voltage to within 1V instead of 3V and that even greater productivity could be obtained if voltage control could be maintained to within 0.5V. The latter level of precise voltage control is not available at the moment for production. It is however, possible to assess production and cost benefits of improved instrumentation or equipment which might justify development of such control. These calculations would have to take into account the capabilities of other approaches, *e.g.* multiwire.

4 Operator training to improve recognition of the importance of process control in determining productivity. Although the submerged-arc process is complex it is evident that general principles can be applied which enable productivity to be improved. Application of process selection techniques to individual production can be assessed by attention to two main aspects: generation of procedural information, coupled with definition of production objectives and capabilities of production plant.

Heat input

Many submerged-arc procedures specify a maximum heat input and it becomes necessary to relate heat input to other parameters. Heat input is defined as the electrical energy supplied to a unit length of weld seam in kJ/mm. This is:

$$E(kJ/mm) = \frac{\text{Voltage} \times \text{current}}{1000 \times \text{speed}(\text{mm/sec})}$$

The heat input necessary to maintain the optimum cross section of weld is independent of welding speed, Fig.5.11*a*, for two sided welding of two different thicknesses of steel plate using a single wire. For multiwire welding heat input can be taken as the sum of the heat input from the individual wires. Again, providing that plate thickness, edge preparation, wire arrangement and flux are the same the heat input required is independent of welding speed, Fig.5.11*b*.

100

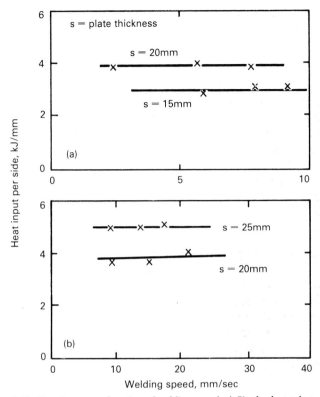

5.11 *Heat input as a function of welding speed:* a) *Single electrode, one run each side;* b) *Three electrodes, one run each side.*

Although the above is an approximation it is a useful relationship to bear in mind when working out welding procedures. As a 'rule of thumb' it indicates that heat input in kJ/mm per side is approximately 1/5th of plate thickness in millimetres. In the example in Fig.5.11 *a* for two sided welding of 20mm thickness plate required heat input per side is 4 kJ/mm.

Further reading

Literature covering the submerged-arc process is extensive and it would be unrealistic to attempt to prepare a full scale bibliography for this book. The following list is intended to act as a starting point for further reading.

AMERICAN WELDING SOCIETY: 'Welding handbook'. 7th edition, Vol.2, Chapter 6, 'Submerged-arc welding', 189-223, Miami, AWS, 1978.

AMERICAN SOCIETY FOR METALS: 'Metals handbook'. 9th edition, Vol.6, 114-152, Metals Park, ASM, 1983.

LINCOLN ELECTRIC COMPANY: 'Procedure handbook of arc welding'. 12th edition, Section 5, 'The submerged-arc process', Cleveland, Lincoln, 1973.

DAVIS L and DAVIS M L E: 'An introduction to welding fluxes for mild and low alloy steels'. Publ The Welding Institute, Abington, Cambridge, 1981.

UNION CARBIDE CORP: 'Submerged-arc welding handbook'. Publ Union Carbide Corp, New York, 1974.

MARTIN L F: 'Fluxes, solders and welding alloys'. Publ Noyes Data Corporation, New Jersey, 1973.

COOK, G E, RANDALL M D, SHEPARD M E and YIZHANG L: 'Adaptive submerged arc welding control'. 'Advances in welding science and technology.' Proc Int Conf, Gatlinburg, TN, USA, 18-22 May 1986. Ed S A David, publ Metals Park, OH 44073, USA, ASM International, 1986. ISBN 0 87170 245 2.

CHANDEL R S, BALA S R and MALIK L: 'Effect of submerged-arc process variables'. *Welding and Metal Fabrication* 1987 **55** (6) 302-304.

BAILEY N: 'High quality submerged-arc welding with metal powders'. 'Improved welding productivity with modern steels'. Proc publ Luxembourg L-2985, Commission of the European Communities, 1986.

KENNEDY N A: 'Narrow-gap submerged-arc welding of steel — a technical review'. *Metal Construction* 1986 **18** (11) 687-691 and (12) 765-769.

TURKDOGAN E T: 'Fluxes for submerged-arc welding'. 'Physicochemical properties of molten slags and glasses'. Publ The Metals Society, London, 1983. ISBN 0 904357 54 6.

FRASER R, McLEAN A, WEBSTER D J and TAYLOR D S: 'High deposition rate submerged-arc welding for critical applications'. 'Offshore welded structures'. Proc 2nd Int Conf, London, 16-18 November 1982. Publ The Welding Institute, Abington, Cambridge, 1983. ISBN 0 85300 168 5.

KEELER T and GARLAND J G: 'How SA welding adapted to the offshore challenge. Part 2'. *Welding and Metal Fabrication* 1983 **51** (4) 193, 195-199.

DAVIS M L E and BAILEY N: 'Properties of submerged-arc fluxes — a fundamental study'. *Metal Construction* 1982 **14** (4) 202-209.

BARCLAY J R and JORDAN M F: 'Future developments in submerged-arc welding: high strength weld metals'. 'Welding in the eighties.' Proc publ Milsons Point, NSW 2061, Australia, Australian Welding Institute, 1980.

INDEX